# the new
# idea factory

## expanding technology companies
## with university intellectual capital

CLIFFORD M. GROSS

UWE REISCHL

PAUL ABERCROMBIE

**Battelle Press**

Columbus • Richland

*HD*
*53*
*· G 75*
*2000*

## Disclaimer

Selecting, licensing, developing, and transferring technology is difficult and time consuming and can be a very costly, long-term investment. This book is an introduction to the sourcing and transferring of university-based technology to small companies. Individual results of these types of efforts may vary considerably. Also, the authors strongly recommend that expert, professional advice be obtained prior to consummating any technology license or transfer. This book is not intended to provide such advice.

Library of Congress Cataloging-in-Publication Data

Gross, Clifford M.
　The new idea factory : expanding technology companies with university
　intellectual capital / by Clifford M. Gross, Uwe Reischl, and Paul Abercrombie.
　　p. cm.
　Includes bibliographical references and index
　ISBN 1-57477-090-X (softcover : alk. paper)
　　1. Intellectual capital—United States. 2. High technology industries—
　United States. 3. Industry and education—United States. 4. Technology
　transfer—United States. I. Reischl, Uwe, 1945–　. II. Abercrombie,
　Paul, 1968–　. III. Title.

HD53.G75 2000
658–dc21

00-027204

Printed in the United States of America

Copyright © 2000 Battelle Memorial Institute

*Ordering Information*
Battelle Press
505 King Avenue
Columbus, OH 43201-2693 USA
Phone: 1-800-451-3543 or 614-424-6393
Fax: 614-424-3819
E-mail: press@battelle.org
Website: www.battelle.org/bookstore

# contents

# tables and figures

## *Tables*

## *Figures*

# preface

A few years ago, when I was a professor at the University of South Florida's (USF's) College of Public Health, a colleague and friend, Dr. George Newkome (who serves as Vice President for Research at USF), asked me to take a look at how the university was marketing its intellectual properties and to offer suggestions for generating a greater return from the university's research work and investment.

Before I made any recommendations, I wanted to see how other large established institutions managed their patent portfolios. So, I tapped into the Association of University Technology Managers, which publishes a report on research expenditures for the nation's top universities. I looked at how much revenue these schools generate from licensing technologies produced by their scientific research efforts, and I did some calculations. Then, armed with this data, I began to evaluate these university R&D machines from a financial perspective, looking at the money earned from licensing patented technologies as a return on equity.

In short, I looked at them as businesses—idea factories, if you will.

I found that the top universities and medical centers spent $24.4 billion on research and development in 1997 and produced 11,784 new invention disclosures and 3,224 patents—or about 200 new technologies each week. Those top schools produced the overwhelming majority of all university-sponsored research in the

United States. Plus, I learned that more than 70 percent of these patented technologies went unlicensed, which seemed like a tremendous waste of the human spirit, potential and innovation known as "intellectual capital."

Next, I calculated the licensing revenue generated by each university. With some notable exceptions, all spent far more money creating technologies than they ever earned from licensing deals cumulatively. Actually, the return on equity (ROE) was about 1.5 percent before expenses. Not much of a way to run a railroad.

The actual product potential of those 11,784 invention disclosures is nearly impossible to estimate—without an efficient market, you really can't know the answer. It's a bit like asking how many homes would be sold if there were no such things as mortgages. Some cash sales would take place, of course, but mortgages are critical to the creation of an efficient real estate market.

I also found in my research that many schools have a director of patents and licensing or some similar professional who decides which technological innovations should be nurtured as potentially patentable products. While these folks may have good eyes for patentable products, this sort of "sole arbiter" process is not terribly efficient.

The bottom line? Universities produce vast amounts of technology that never get to market. Chief among the reasons for this is the focus on large company licensees, coupled with the mission to look for big, obvious, knock-it-out-of-the-park-type ideas. Unfortunately, many of these would-be home runs turn into strikeouts because of the lack of an efficient mechanism for getting technologies to market. Sometimes they fail for lack of financing. In some cases, due to the speed in which ideas can be converted to intellectual capital, there simply *is* no market—yet.

As an example, fiber optic cable's curious ability to carry light had been around for some time before researchers at Corning Inc., in the 1970s, figured out that the gossamer filaments of glass might also be useful in telecommunications. Today, nearly all U.S. long distance phone call traffic travels at the speed of light through fiber optic cables. And fresh innovations promise to pump up its information carrying capacity further.

Moreover, many university patent officers look only to the biggest companies to shop their technology, assuming that these corporate behemoths are best able to bring their school's technology to market. At first glance, this makes sense. But it overlooks the fact that large national and multinational companies often have big, well-funded internal research development divisions of their own. Big companies spend billions each year on internal R&D—they often are funding 100 or more research projects of their own at any given time. And that discourages them from looking at largely undeveloped technologies that were created somewhere else.

But smaller companies are another matter. Small technology companies have tiny or non-existent R&D budgets, but they are ravenous for new technologies that they can turn into products. Fresh, proprietary products are the key to these companies' survival, yet many can't afford to do their own development. In short, they have the will but not the wallet.

While small companies often feel that they must develop all of their own technology, few have effective R&D capabilities. Partnering with someone with similar R&D interests makes enormous sense to these companies. Where can a small company find a nearly unlimited supply of cutting-edge technology, much of which can be turned into profitable products? Why, the idea factories—the universities.

The need became apparent for a new model for technology transfer—one that merges the needs of the technology suppliers (the universities) with small-cap companies that are hungry for new technologies to survive and grow, but are usually short of cash. A bridge needed to be developed to enable both universities and companies to get together more efficiently.

Key to this novel model of technology transfer is an innovative method of financing the acquisition of technology, in which small-cap customers can acquire various technologies from a technology merchant in exchange for stock, often the only currency a small company has. Besides offering a mechanism for cash-strapped companies to acquire potentially breakthrough proprietary technology, this new method of technology transfer is especially attractive to universities because, unlike traditional methods in which companies acquire licenses to technologies, the university retains ownership of its own technology.

Additionally, with this model, universities receive 100 percent of any royalties. This is a revolutionary concept for schools and their researchers, who, through traditional technology transfer arrangements, typically receive a fraction of revenues earned from their inventions.

*The New Idea Factory* is aimed at helping entrepreneurs, senior managers of technology companies, technology transfer professionals, the inventors and investment bankers more easily identify and transfer technologies that will contribute to the growth of the economy in the coming decade. Also, it describes a novel strategy to forge alliances with these "idea factories" to turn technologies into profitable products. We call this U2B™.

In particular, we hope that this book will help the thousands of small-cap and startup companies play—and win—in the ever-changing global marketplace.

As technology and technology companies dominate the world's marketplace, companies need more than ever to be able to compete on the intellectual capital playing field. *The New Idea Factory* is the guide for leveraging the brainpower capital found in universities and government laboratories and for creating wealth from the strength of these new ideas.

—CLIFFORD M. GROSS

*In memory of Rose Mark*

# acknowledgments

The gratitude we feel toward those who have helped us to review and test some of our ideas looms large. We are especially grateful to Dr. George Newkome and Dr. Kwabena Gyimah-Brempong of the University of South Florida for starting us down the path of extracting value from intellectual capital; to Ken Preston for helping us better understand the many opportunities and concerns regarding university licensing; to Drs. Frank Hong and Eugene Lee for their assistance and confidence in helping to finance our early efforts; to Sam Reiber for his steadfast support, legal perspective and the sense of humor and camaraderie that he has helped to foster; to Carl Nisser for his friendship and global perspective on technology merchandising; to John Lloyd for his faithful help in building our data systems; to Nina Siegler at Johns Hopkins, Dr. John Mott at Los Alamos National Laboratory, Michael Comella at Cornell University, Dr. Michael Rollor at the University of Maryland, Dr. Linda Brinkley at the University of Memphis and Chris Moulding at Caltech for having the enthusiasm to utilize a new technology transfer model. Thanks to Diane Wiles, Susan Carr, Bea Weaver, Allison Gonczy and Aileen Reischl for help with the manuscript preparation and research, to Arthur and June Chapnik, Florence Hassell, Debbie Radsick, Seth Frielich, Fergus Norton, Dr. Albert Anthony, Dr. Gerald Krueger and Dr. Wolfram Weinsheimer for their combined efforts at trying new methods to inspire small public companies to build bridges with university research centers;

to our European Advisory Council members Lord Chesham, Dr. Jonas Lonnroth, Bo Hjelt, Professor Marcel Crochet and Lupold von Wedel for their willingness to teach us about the unique opportunities in Europe regarding technology transfer and methods to address them; to Dr. Stuart Brooks for his practical sensibility regarding medical issues; to Larry Oremland for his contribution to our understanding of the spirit, as well as the subtleties, of intellectual property; to Dave Ritzert, Warren Hampton, Jigna D. Desai and Dick Dobkin of Ernst and Young for their insight into new approaches for valuing intellectual capital; to Carole Mason for her ability to account for the ephemeral in concrete terms; to George Getz and Bill Foster for their early assistance and guidance; to Jay Kaplowitz, Henry Nisser, Arthur Marcus, Eric Dixon and Andy Russell at Gersten, Savage and Kaplowitz and Jeff Ellentuck at Reed, Smith, Shaw & McClay for their legal and structural insight; and to Max Rockwell and Owen May at May Davis Group for having the vision to help build and capitalize a new organization for acquiring, developing and transferring university inventions to the marketplace. A special thanks is also due to Beatrice Bryan and Diane Hoffman for helping us to utilize The Association of University Technology Managers detailed data on university research. Without this formidable team we would have done very little. Their combined contributions have enabled us to connect our ideas and begin to test their veracity under load.

CLIFFORD M. GROSS
UWE REISCHL
PAUL ABERCROMBIE

# great thinking machines— the new idea factories

1

> *The real act of discovery consists
> not in finding new lands but
> in seeing with new eyes.*
> —MARCEL PROUST

EHIND EVERY GREAT PRODUCT OR PROCESS IS AN IDEA—SOME
person (or persons) using brainpower to create what didn't
exist before, whether it's an entirely new way to diagnose
disease or a fresh twist for facilitating e-commerce. With the speed of
technological advances and information being deployed at ever-
faster rates, creating new products and services is crucial for companies that increasingly depend on significant innovations for survival.

Innovation is the engine running global economies, and it also
drives people to improve their living standards. People crave innovation—it is the creative spark that makes humans more than mere
carbon-based bipeds. At no time in our history has innovation been
as important as it is today. Indeed, finding solutions to overpopula-
tion, disease, pollution and the rest of the world's problems is paramount. To solve these problems will require more than incremental
improvements—we will need the strength of big ideas.

Information and knowledge are quickly replacing bricks and mortar as the building blocks for future survival and fulfillment. More than ever before, producers and synthesizers of knowledge are going to be the foundation for the development of our economy and, indeed, our entire society. And where are such purveyors of knowledge likely to be found? At universities, the idea factories of the 21st century.

The faster we link the capabilities of our innovative and creative universities to the marketplace, the more we invigorate the economy with new ideas. That's why the new idea factories will be the primary engine for economic development and growth in the 21st century.

All of this will lead to more than just new entertainment gadgetry. University innovation can lead to new products and services that can save lives, reduce suffering and improve the quality of life worldwide. Successful technology transfers have led to everything from new methods for the early diagnosis of cancer to improvements in highway guardrails.

If we are to nurture and harness technology growth and development in the next century, we must foster direct linkage and interaction between our new idea factories and the commercial sector.

We are in the midst of an information revolution that began with the cold war's race into space and was accelerated with Intel Corp.'s development of the microchip. This revolution is still in its infancy, and it is picking up speed.

In attempting to improve the standard of living throughout the U.S. and the world, it will be necessary to increase technological progress nationally as well as globally. Part of moving forward means using what we have—and currently we make poor use of our university "knowledge factories." Just imagine how much faster our growth and development would be if we fully used all of the ideas and knowledge created in our colleges and universities every single day.

The model described in this book involves constructing a bridge between university inventions and technology companies that are hungry for them. This is a connection that, if nurtured, can bring new ideas to the marketplace, create wealth and ultimately improve our standard of living.

This type of partnership will work wherever there is fertile ground for new ideas and innovations. Different and diverse ideas and processes must be brought together to create a social and economic environment of openness and possibility.

If a successful real estate deal hinges on "location, location, location," a technology success story could be said to rely upon "diversity, diversity, diversity." Cultural, ethnic and idea diversity provides a rich intellectual "gene pool," and it is this that leads to tolerance of new ideas and to the realization that there is always more than one way to get something done.

Many people have achieved success and wealth. Many of these successes have been made possible because those who can create value by introducing new ideas are rewarded. When we listen to the rhythm of a culture, we should hear the message that ideas and contributions matter!

This is an important message for everyone. But when that same message is delivered to more than 250 million people in the United States, the result can be a collective level of energy that is truly unique in the world. In the U.S., people are encouraged to look at things in unique and different ways—to tolerate ambiguity.

Consider California and New York, two states that generate huge numbers of new ideas every day. Many international companies have plugged into this creativity by establishing their design and development headquarters in these two states, especially in California. Why? Because California offers global diversity, and diversity can inspire innovation. Many ideas that develop there often spread around the globe. In New York, a so-called "Silicon Alley" has sprung up in the lofts of the garment center, Tribeca and the theater district—nurturing Internet startups and their IPOs, with their unique and not-for-the-squeamish business models.

There is, however, no geographic monopoly on inventiveness. In the United States in 1998, Idaho produced the most patent applications per 100,000 residents (Table 1-1).

Although lacking in traditions that are centuries old, as well as a rich and complex history, the United States is very open to the rapid development of new ideas. We may well be one of the most creative countries on earth, inasmuch as more intellectual capital is created in the United States than anywhere else in the world. Why?

Renowned for attracting outstanding students from around the world (in part because of well-funded research), U.S. graduate schools also attract scholars and researchers from every corner of the globe. Thankfully, many of the students remain in the United States when their education is completed, contributing to this country's diversity. In the future, as in the past, this pool will be an important source of new scholars and scientists.

U.S. public schools may not turn out students with the highest average scores in math and science, but they do turn out students who bridge the gap between what they know and what is needed in the creative pursuit of knowledge. The U.S. culture seems to produce students who can evolve to investigators, innovators and entrepreneurs—i.e., creative thinkers.

Although there is room for significant improvement, we must be doing some things right—we have an explosive economy, are home to 36 of the 50 largest corporations in the world and have more Nobel laureates than all other countries combined. We have a highly diversified intellectual capital foundation.

## Table 1-1. The most inventive states in the U.S.

Ranking of states in the U.S. by patent applications filed per 100,000 residents in 1998

| Rank | State | Patent Applications Filed | Patent Applications per 100,000 Residents |
|---|---|---|---|
| 1. | Idaho | 2195 | 178.6 |
| 2. | Massachusetts | 6467 | 105.2 |
| 3. | Vermont | 567 | 96.0 |
| 4. | California | 31135 | 95.3 |
| 5. | Minnesota | 4302 | 91.0 |
| 6. | New Hampshire | 1052 | 88.8 |
| 7. | Connecticut | 2905 | 88.7 |
| 8. | Delaware | 651 | 87.5 |
| 9. | New Jersey | 6854 | 84.5 |
| 10. | Colorado | 3153 | 79.4 |
| 11. | Oregon | 2378 | 72.5 |
| 12. | Washington | 3325 | 58.4 |
| 13. | Arizona | 2670 | 57.2 |
| 14. | New York | 10289 | 56.6 |
| 15. | Michigan | 5528 | 56.3 |
| 16. | Utah | 1161 | 55.3 |
| 17. | Texas | 10588 | 53.6 |
| 18. | Wisconsin | 2794 | 53.5 |
| 19. | Illinois | 6385 | 53.0 |
| 20. | Rhode Island | 517 | 52.3 |
| 21. | Pennsylvania | 137 | 51.1 |
| 22. | Maryland | 2543 | 49.5 |
| 23. | Ohio | 5508 | 49.1 |
| 24. | North Carolina | 3165 | 41.9 |
| 25. | Indiana | 2406 | 40.8 |
| 26. | Iowa | 1086 | 37.9 |
| 27. | Nevada | 620 | 35.5 |
| 28. | New Mexico | 568 | 32.7 |
| 29. | Florida | 4838 | 32.4 |
| 30. | Georgia | 2396 | 31.4 |
| 31. | Virginia | 1914 | 28.2 |
| 32. | Missouri | 1488 | 27.4 |
| 33. | Tennessee | 1480 | 27.3 |
| 34. | Kansas | 672 | 25.6 |
| 35. | South Carolina | 941 | 24.5 |
| 36. | Oklahoma | 792 | 23.7 |
| 37. | Montana | 204 | 23.2 |
| 38. | Nebraska | 338 | 20.3 |
| 39. | South Dakota | 149 | 20.2 |
| 39. | Maine | 251 | 20.2 |
| 39. | Louisiana | 881 | 20.2 |
| 42. | Kentucky | 698 | 17.7 |
| 43. | Wyoming | 83 | 17.3 |
| 44. | North Dakota | 107 | 16.8 |
| 45. | Alaska | 102 | 16.6 |
| 46. | Alabama | 694 | 15.9 |
| 47. | West Virginia | 251 | 13.9 |
| 48. | Hawaii | 163 | 13.7 |
| 49. | Arkansas | 326 | 12.8 |
| 50. | Mississippi | 280 | 10.2 |

***Sources:*** 1998 Annual Report of Commissioner of Patents and Trademarks and 1998 population estimates by U.S. Bureau of the Census. Includes filings from the 50 states plus the District of Columbia, Puerto Rico, Virgin Islands, and U.S. Pacific Islands. District of Columbia residents filed 116 applications, or 22.2 applications per 100,000 residents.    © 1999 Intellectual Property Owners Association

## THE WRIGHT BROTHERS

A good illustration of the way theoretical and practical approaches affect scientific and technological innovation is the development of the powered airplane.

In Europe, Otto Lilienthal was doing aviation research at around the same time that the Wright brothers were developing their flying machine in the U.S. In fact, Lilienthal had started doing some excellent work a number of years before the Wright brothers, although he focused on gliders rather than powered machines. There was, however, a great difference in the way Lilienthal and the Wrights did their work.

Lilienthal studied aerodynamics. He studied birds, he conducted experiments and he looked into the best shape of wings. Perhaps 15 or 20 years before the Wright brothers' first flight, Lilienthal was flying gliders. He would jump off a mountain using wings he had designed and built. He published books and scientific papers—he was a scientist, the leader in the field for all of Europe, an expert in the field of aviation.

He did not, however, invent the powered airplane. That accomplishment belongs to the Wright brothers, even though Lilienthal was the preeminent aviation expert.

What gave the Wright brothers the advantage over Lilienthal? In the beginning, the Wright brothers weren't really interested in aviation at all. They were bicycle builders, and in those days bicycles were high tech. You might say the Wright brothers were like the Silicon Valley innovators of today. High-tech bicycles, the use of new material, the use of wire-spoked wheels, were all really cutting edge in those days.

Their obvious next step as bicycle builders was the motorcycle, a motorized version of what they were already doing. It was the motorcycle's power plant that inspired the idea of putting a similar motor on a flying machine.

The brothers had to learn about aviation, and they studied the ideas of Lilienthal. The Wright brothers did plenty of research, and they presented their findings at scientific conferences, just as we do today.

The Wright brothers were hands-on people, while Lilienthal was more of an academic. Lilienthal was accepted; the Wright brothers were outsiders. But they were able to apply some of Lilienthal's theories on wings to the propeller, which was the important element lacking at the time. The airplane propeller is very much like a spinning wing. The brothers realized the parallelism and did breakthrough work. However, the decisive (and patented) element was the adjustable aileron. With this technology, it was possible to change in mid-flight the shape of a plane's wings—just like the birds—and in doing so achieve controllable and stable flight.

The Wright brothers may not have been great theoreticians, but they were practical. Lilienthal may have been the academically recognized scientist, but the Wright brothers were the epitome of American can-do. Their work became the foundation of our modern aviation industry. Today, millions of people rely on commercial aircraft for business and pleasure.

## THE PUSH/PULL OF MARKETS AND TECHNOLOGY

Technology solutions are driven by need. And solutions in turn drive the technology further. Consider the many possibilities that emerge from partnerships between researchers and industry in relationship to the technology marketplace.

What pulls technology through the marketplace? Change—changes in environmental requirements, legislation, the need for clean water and clean air, better and easier to use software and greater speed and power of microprocessors to quench our thirst for instantaneous, global communication.

But technology gets pushed, as well. For example, a small company may drive technology as it establishes itself in its particular technology niche and demonstrates a paradigm shift, i.e., Netscape or Amazon.com.

We end up with a push/pull scenario. The market demands (pulls) the best technology available, while companies push specific technologies they need and want to develop.

What if companies see the need for a more competitive product, but don't have the resources or the time to develop it? The only real alternative is to build technology transfer relationships with major universities or government labs that have the capabilities to advance such technologies. Technology transfer is an effective tool for achieving technology parity with larger, more powerful corporations, although it is underutilized by most small public companies. Technology transfer is an important tool to help small companies make the weakness of their own technology development capabilities irrelevant.

In today's marketplace, speed not only determines which products get to the market first; it can also be the determining factor in who gets to establish the standards of any given market segment.

The computer business provides a good illustration. The company that gets to market first may well be able to set the standards for the entire market segment. The establishment of defective standards is not an unheard-of phenomenon in this scenario. Perhaps the best example is the computer keyboard. Many critics believe that the contemporary keyboard is laid out poorly. The design is more the result of tradition than of good ergonomics. Designs are often determined by who gets there first, who brands first or who better knows how to leverage distribution channels.

## FADS AND TRENDS

Today, many products result from interesting or unusual technologies rather than from actual need. The need, however, may become apparent later. The fax machine might qualify under this description—the early fax machines weren't necessary because they did not deliver any more information than the U.S. mail. What they did do was deliver it faster—a lot faster. This acceleration of information exchange leads directly to enhanced productivity, which is the key to enhancing economic vitality in general.

Creating need—crossing the line from fad to trend—can be the point where companies actually cash in on a technology. Many companies are good at getting in on fads through innovation, but less effective at cashing in because they don't have the necessary delivery system.

What really pushes technology is an innovation that makes a real improvement and thereby defines a new niche. Consumer use of a technology is the ultimate pull. Windows 2.0 was a push; Windows 98 was pull, pull, pull.

Prior to 2.0, Windows wasn't even a dominant platform. "Pull" happened when Microsoft was able to say, "All right, we have a huge installed basis looking for applications, looking for improved speed and performance." Now Windows was being pulled by the vortex of a rapidly expanding market.

## TOOL FOR CEOS

Technology transfer, or turning university-based scientific innovation into products and processes, is a powerful tool for the CEOs of small and mid-sized public companies. Technology transfer can help maintain equilibrium between market pull and technology push for particular market niches. It can be like a technology "spice rack," where you can take the spice that you need to enliven the sauce a little bit. Or more intensely, it can be a hot pepper, like turning

Mosaic into the Netscape browser. Technology transfer is a tool that transforms science into useful products, including the potential for profits not only for the universities who create a technology, but also for the companies that acquire and move the technology to the marketplace.

# technology trends for the 21<sup>st</sup> century

# technology trends for the 21st century

*Vision is the art of seeing things invisible*

— JONATHON SWIFT

THE HISTORY OF HUMANKIND IS THE HISTORY OF TECHNOLOGY—THE axe, the wheel, the bow and arrow. While these tools may seem primitive now, they were high tech for their day. Some of our earliest tools, discovered in Hadar, Ethiopia, are referred to as choppers and date from 1.5 to 2.6 million years ago. From the cognitive to the mechanical, our ability to make tools helps to define us as a species (Gross, 1996).

Most technological innovations came about in response to a particular problem. The wheel revolutionized transportation of people and goods, the bow and arrow helped people hunt for food or protect themselves, and the telephone helped individuals communicate at a distance. Interestingly, the more developed a technology becomes, the less it costs. A case in point is the price of a three-minute call between New York and London (Figure 2-1).

Modern technology has spawned many useful products, a growing portion of which has been relegated to entertainment (CD-ROM

## FIGURE 2-1
Cost of a three-minute telephone call from New York to London

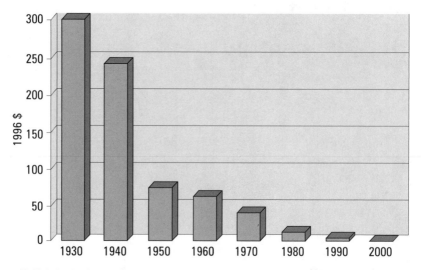

Modified after data from *The Economist,* October 18, 1997, and Ungson and Trudel, 1998.

music players and Internet chat rooms, for example). Such technologies can be used to help satisfy basic human needs, such as communication.

Forecasting which technologies will catch fire in the 21st century is like trying to read tea leaves. But there are global trends of such strength and importance that they should continue well into 2000.

The following basic principles can be helpful in deciding which technologies are worthy of investment:

◆ Technologies that represent a major reform or breakthrough, in other words, a major leap rather than an incremental step.

If it's not a big technological advancement, chances are it will be overtaken by competing technologies before it can be fully developed and marketed. Also, a leap is necessary to maintain an intellectual capital-potential energy advantage.

◆ Technologies that have a global market. Technologies of regional importance, no matter how big a technological jump forward, typically have a smaller potential for profits. Furthermore, technologies of global impact assume an expansive market with room for multiple players—at least early on.

◆ Technologies that are socially responsible.

The United States has been the world's leader in technological entrepreneurship. Technology has become a global commodity that is developed, traded, sold, and marketed in every corner of the world. Never before has the world seen so much international trade and competition for global markets. This trend will undoubtedly continue during the 21st century. To survive and grow in this competitive environment, companies must find relevant new technologies, acquire them and put them to work. They must use the new technologies to improve efficiency, reduce waste and energy consumption and create new products and services that are attractive, easy to use and fun. The globalization of technology not only challenges the survival of today's companies, but offers exciting opportunities for growth throughout the world (Battelle, 1999).

To take advantage of future opportunities, technology trends should be identified early and incorporated into the strategic business plans of the entrepreneurial companies. A team of scientists and engineers at Battelle recently published a list of technological challenges that they believe will confront industry in the 21st century (Battelle, 1999).

According to the Battelle team, the major challenges and opportunities for the 21st century will include the following:

◆ **Renewing the public infrastructure.** *In the developed countries around the world, the public infrastructure that provides transportation, bridges, water and sewage removal is deteriorating with age. Costs of major new projects will be huge. New materials and new construction methods will be required to renew the infrastructure with limited public funding. New infrastructure needs will include traffic control and management systems that reduce travel times and efficient and practical mass transit systems* (Battelle 1999). To these ends, engineers are designing revolutionary materials for use in constructing everything from stronger, longer-lasting skyscrapers to safer, faster airplanes.

Just as the human genome project is helping us figure out what makes our bodies tick and how to fix or improve them (i.e., preventing congenital deformations), advances in computer technology are allowing engineers to design and create all sorts of novel building materials molecular brick by molecular brick (nanotechnology). For example, an innovative carbon-fiber technology promises to help in the creation of a new category of reinforcing bars for concrete structures with a life span of possibly more than triple that of traditional concrete structures reinforced with steel.

Along with longer life, many of these new materials will be "smart." That is, they will be able to give off warning signals of various kinds when they detect excessive stress and perhaps even mobilize themselves to minimize stress concentrations (i.e., heal the fracture).

◆ **Developing mobile, high-density energy sources.** People are becoming increasingly mobile, and they want easy and rapid communications, which require highly mobile energy sources. One of the biggest complaints people have about current notebook computer technology is poor battery life. Companies and homeowners also need more flexible power sources. In the future, many more cars and homes will operate on alternative fuel systems.

◆ **Protecting the environment.** Much of the economic growth in the past *was fueled by easy exploitation of our natural resources. Many of these resources are now largely tapped out or damaged, so that further growth will come from the smart management of the remaining resources and our ability to use alternative resources. We need new technologies to provide for the long-term sustainability of our natural environment including air and water. We need to find ways to increase the efficiency of energy production and conversion* (Battelle 1999).

Technologies that improve the food and water supply will play an especially important role in the next millennium. Near-term exponential population growth for some countries (India's population just topped the 1 billion mark) will put great strains on global supplies of clean water. In turn, as supplies of natural sources of water become ever scarcer, we're being forced to find alternative methods for finding new supplies of fresh water or methods of producing it.

In many places in the world, bottled water is more expensive than gasoline, and it's going to get more expensive. We'll have to find new ways to recycle water with fast, inexpensive purifying technologies. The need for cost-effective desalination will become paramount. Additionally, technologies that help remove toxic substances from the air, land and water will become increasingly important.

Interestingly, as the global climate heats up, increased rainfall will follow. This increase in precipitation will help nourish many disease-causing organisms that require water for survival. In addition, increased rainfall and flooding will increase their dissemination, and higher temperatures will enhance their survivability. This would apply to bacteria such as *Salmonella* and *Shigella*, viruses and protozoa such as *Giardia* and *Cryptosporidium* (Martins, 1999). Therefore the need to be able to rapidly and accurately detect and eradicate water-borne contaminants will continue to expand.

No doubt nations around the world will become more eco-conscious as we realize that to survive we must preserve the fragile balance between nature and our technologies. As computer power increases and we improve our ability to create realistic mathematical models of our environment, we will be able to predict the environmental consequences of our production and consumption.

This trend will, in turn, force us to further refine alternative methods for capturing and storing energy and to develop fresh technologies to replace or improve the efficiency of our fossil fuel energy sources.

◆ **Providing personal security.** *In the wake of car hijackings, gang violence and terrorist bombings in major cities, many people feel threatened. New technologies are needed that can make the shift from national security—protecting nations from invading armies or missiles—to personal and community security. We will need new technology-based methods or systems to better protect us from crime and terrorism at home, at work and in our schools (Battelle 1999).*

◆ **Providing human interfaces.** *Interfaces are the intermediaries between people and machines that allow us to more easily use* *technology. . . . As more complicated technologies flood the home and the workplace, consumers will demand interfaces that go the next step beyond "user friendly." Tomorrow's successful technology-based products and services will have to be pleasing to our senses, more or less intuitively obvious, safe to use and, most of all, fun* (Battelle 1999).

The way we use computers and other tools will change radically in the coming century. The trend toward improving the way we use tools is nothing new. For instance, improving the way a hammer fits in your hand not only allows more efficient work, but also helps mitigate potential stresses and injuries created by repetitive motion. In short, it's a better interface between human and task. At its best we call this "creating a transparent interface." When Black & Decker improved the interface design of its DeWalt cordless drill line, the drill not only felt more comfortable to the user, it rapidly became the global market leader.

Such ergonomic considerations apply also to the way we interface—and even communicate—with such complex, increasingly ubiquitous tools as computers and telephones. Over time, cumbersome computer key punch cards and bulky hand controls gave way to touch-type keyboards and the mouse, which has itself given way to the touch-pad mouse. Improvements in voice-recognition technology are making speech a viable method of operating any number of machines, from telephones and computers to air conditioning systems and automobiles. Technology interfaces will change to meet demand. Because most people find it easier to speak than to type, voice recognition (or ultimately thought recognition) technologies will probably grow in popularity.

And technologies that "humanize" machinery or otherwise counteract the coldness of communication between people and machines will probably grow in importance in the current century. Scientists at the Jet Propulsion Laboratory have developed a technology to create photo-realistic, digitized people that promise to improve the way we interact with computers, the Internet and ourselves. This once top-secret technology breaks down videotaped images and audio recordings of a person into digitized facial gestures and phonemes, the smallest units of speech. With these building blocks, it is possible that a person's face and voice may be reconstructed so that a computerized countenance can be directed to say anything. Picture an animated Max Headroom, but as realistic as a live TV news broadcaster.

Such technology may humanize surfing the Internet. Or the way companies or even government agencies present themselves to the public via the Internet. Among the potential uses for these "malleable mugs" are digital customer service representatives for all types of organizations. Or they could be used as tireless virtual teachers or news and information broadcasters. These photo-realistic folks could either respond to spoken questions with prerecorded answers or operate as a façade for an artificial intelligence agent powering text-to-speech algorithms.

Assembling faces and voices in any combination would allow clients to mix and match, in effect giving companies the ability to construct the ideal representative for any occasion—from a kindly physician delivering advice over a medical Web site to bucolic service agents representing organizations with high-volume customer interactions.

◆ **Developing personalized consumer products.** *Consumers are increasingly better informed and harder to please. In the future, they will want products that satisfy their own tastes rather than whatever stores present. Products will have to be almost as varied as individual customers. This market force will require companies to be even more consumer-driven in designing and marketing their products. Sensors, controls and computers must achieve highly flexible manufacturing of customized products, i.e., mass customization* (Battelle 1999).

The way we use machines will involve less obvious physical manipulation. This "naturalizing" force of technology interfaces increases in parallel with the sophistication of the underlying technology. In other words, the better the technology, the easier it is to use.

Over time, sun dials and sextants have evolved into slim, accurate electronic watches and global positioning systems no bigger than a pack of cards, on their way to becoming as small as the watch or smaller. Indeed, the interface between our technology and us is rapidly becoming quite literally interwoven. We "wear" pagers and telephones, which are growing ever smaller and more efficient. Soon our technology will become as inextricably stitched into our lives as our clothes. Our telecommunications gear might be quite literally woven into the shirts we wear.

What's more, future interface technologies may allow people to communicate with personal knowledge assistants by thought alone. Perhaps electro-chemical signals given off by the brain will be harnessed to create seemingly clairvoyant interfaces between human and machine.

This merging of man and machine has been going on ever since people began making tools (i.e., a quiver worn on the back as a handy way to tote arrows).

◆ **Converging technologies in the home.** *Increasingly, the home will be a place to work, shop, get an education, be entertained and be safe. The biggest technological change occurring in our personal lives will be the convergence of telecommunications, entertainment and networking in the home. The technology challenge is how to empower and to protect individuals in their own homes* (Battelle 1999).

Along with snappier, more powerful computers comes a vast pool of digitized information flooding the Internet and the airwaves. Most Internet search engines have managed to catalog only a fraction of the ever-expanding universe of Web sites. And the pace of creation of new information is accelerating rapidly.

Like a scene from a popular 1950s horror flick, the Internet's great blob of information grows by the nanosecond. Yet the trick will be transmuting this primordial ooze into something more suitable for human consumption—knowledge. Technology that helps people make sense of this ever-expanding mass of information will be paramount in the 21st century. Sure, the Internet is amazing now, and one can find information on a bewildering array of topics. But the Internet is still in its rapidly growing infancy.

This new global network is growing at a rate not unlike bacteria in a culture (Figure 2-2).

Even when you think you've honed your search words to a razor's edge, you still have to wade through a mountain of information for actual knowledge—the answer to, say, your

FIGURE 2-2

The growth of internet host computers between 1981 and 1998

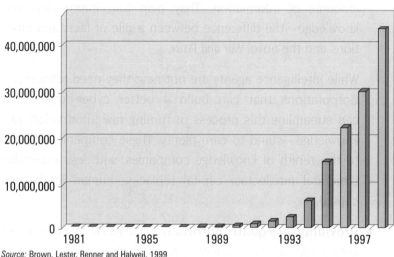

Source: Brown, Lester, Renner and Halweil, 1999

question of how deep to set concrete footings for the pergola you're building in the back yard.

Technology that helps cut through the growing thicket of information to find the right information, and then help make sense of it, will be among the most important—and potentially lucrative—technologies in the coming century. Imagine bits of computer code that act as a kind of virtual butler, who "listens" to your commands, "learns" about your likes and dislikes and then goes to the Internet to find what you're looking for, whether it's a cheap plane ticket to Tulsa or a recipe for tiramisu. Picture a good version of HAL, the cruel, talking computer in Kubrick's 2001.

Powered by advances in artificial intelligence technologies, these information agents increasingly will be able to perform more complex chores. Being able to "think" gives these bits of computer code the ability to organize and synthesize huge amounts of information. They help turn information into knowledge—the difference between a pile of facts and emotions and the novel *War and Peace*.

While intelligence agents are not new, they need refinement. Corporations that can build a better cyber-butler—that can streamline this process of turning raw information into knowledge—stand to earn plenty. These companies will rise to the zenith of knowledge companies, with less expensive structural intellectual capital replacing human intellectual capital.

♦ **Providing adequate nutrition.** *While people in the developed world are becoming more concerned about the nutritional quality of the foods they eat, the rapidly growing population in many poor countries will simply need more food with a high enough nutritional value to sustain a healthy life. Technologies are being developed to engineer natural foods that will be packed with more vitamins, proteins and other nutrients. Other foods will have higher yields, longer shelf life and natural resistance to pests. Packaging techniques will also increase the shelf life of foods, allowing us to store them longer and transport them further* (Battelle 1999). Additionally, farmable land is shrinking at an alarming rate, giving way to new homes, towns and cities. Technology that helps us squeeze more food from less land— or is not dependent on land at all—will grow in importance.

Technology that protects and/or boosts the food supply, and that ensures that it is safe for consumption, is vitally important. To this end, Los Alamos National Laboratory has developed a technology that allows for the almost real-time detection of the presence of pathogenic organisms in food and water. By sampling and identifying a specific component of a target organism's DNA, it may soon be possible to continuously sample perishable foodstuffs to determine the presence of dangerous microorganisms prior to packaging, shipping and consumption.

Technology that cleans or desalinates seawater, rendering it drinkable, will be absolutely vital. The particular importance of water is obvious. There are any number of alternative fuel sources—solar, nuclear, chemical, etc.—to power a lamp or a car. But there is no substitute for water.

◆ **Providing affordable health care.** *In the U.S., market forces are shifting health care from hospitals and HMOs to private homes. Increasing home health care will help contain rising costs, while serving an aging population, and will provide people with the convenience and privacy of taking care of themselves and their loved ones in their own homes. Home health monitors and treatments and linkages to professional care centers present a huge challenge—and an enormous business opportunity for the health care industry* (Battelle 1999).

The human genome mapping project—that is, the worldwide effort to map the entire genetic blueprint for making human beings—will no doubt revolutionize the way we treat ailments of all kinds. The genome project promises to unlock many of the biggest secrets of the human body.

It is estimated that errors in our genes are responsible for more than 3,000 diseases. Furthermore, it is known that altered genes play a part in cancer, heart disease, diabetes and many other common conditions. In complex disorders, genetic alterations can increase a person's risk of developing that disorder. The disease itself results from the interaction of genetic predispositions and environmental factors, including such habits as diet and lifestyle.

By decoding the body's most minute architecture, scientists and doctors may be able to identify and develop cures that have long eluded us. What once seemed like science fiction is on the verge of becoming scientific fact. For instance, doctors may soon be able to look at an individual's DNA and determine whether the person will suffer from such ailments as diabetes or Parkinson's, or whether a person is genetically predisposed to certain types of mental illnesses. Even the act of looking at the DNA will change. At first, doctors will use traditional invasive techniques, such as extracting blood. But soon they will be able to use non-invasive methods, such as spectroscopy to "read" chemical components by measuring how light is reflected by the various tissues of the body.

The genetic diagnosis, treatment and prevention of disease will steadily replace current pharmacological methods and the maturing industry that encompasses them. No longer will doctors treat a patient with drugs to arrest or knock out ailments; instead they'll use genetic medicine to change a person's DNA coding to eliminate the ailment from the person's makeup. "*Genetaceuticals*" will offer cures (i.e., prevention of diseases) or at the very least, mitigate the effects of disease.

For example, doctors may soon be able to determine if you—and any of your existing or potential children—have a genetic predisposition to leukemia. It might be possible through gene therapy to modify the body's genetic coding so that the leukemia never manifests itself. This way, you could obviate a potential problem instead of treating its symptoms.

Also, knowing how a person is made to the last gene may allow scientists to develop replacement parts for transplantation in humans. For example, an individual who suffers from alcoholism and who has developed cirrhosis of the liver may be able to have a new, genetically "fixed" liver grown outside the body that can then be installed. This may sound far-fetched, but scientists are succeeding with similar small-scale pilot experiments right now.

Or picture another scenario: your ear is severed in an accident. So, your doctor calls up your DNA map and orders a replacement ear to be cultivated. In some undetermined amount of time, the replacement part is attached. Scientists already have successfully cloned and grown just such an ear outside a person's body. This same technology might be used to create genetic medicines that can induce limbs and other body parts to regenerate themselves, much as some animals such as certain types of lizards can regrow a severed tail.

Genetic technology might also be used to create break-through cures for such worldwide health problems as AIDS. Billions of dollars are spent each year trying to find a pharmacological magic bullet to kill the AIDS virus. Yet a genetic vaccine may one day render AIDS as easily cured as syphilis, which before the invention of penicillin killed thousands.

And then there is the possibility of solving—at least in part—humankind's most enduring puzzle: how to squeeze more life from life. Innovations in genetic technologies will inspire us to figure out ways to extend life, to repel not just wrinkles and age spots, but death. As you can imagine, the possibilities for benefit or harm to humankind—and the potential financial profits from such technology—are enormous.

Our newfound skill to study genes, however, may become a double-edged sword. For some hereditary diseases, the ability to identify and detect the responsible defective genes has outpaced our ability to do anything about the diseases they cause. Huntington's disease poses such a dilemma. Although a predictive test for high-risk families has been available already for many years, only a small percentage of the persons at risk have decided to be tested. Why? Because there is no cure or prevention for this disease. Undoubtedly, many of the individuals would rather live with the uncertainty than with the knowledge that they will become fatally ill sometime in midlife. And what might happen if a health insurance company or a potential employer learns that an individual is destined to develop the fatal disease?

Because of such concerns, the Human Genome Project devotes about 5 percent of its budget to research aimed at anticipating and resolving the ethical, legal and social issues likely to arise from its research. This is the first time in history that scientists have begun to explore, systematically and at an international level, the potential consequences of their research before a crisis has arisen. With careful attention to these ethical issues, and appropriate safeguards, society can probably enjoy the benefits of the Human Genome Project.

The scientists and engineers at Battelle went further and developed a list of 10 strategic technologies that they believe will characterize the technology landscape of the 21st century. They include the following, in order of importance:

1. Human genome mapping and genetic-based personal identification and diagnostics that will lead to preventive treatment and cures for specific diseases.

2. Computer-based design and manufacturing of new materials at the molecular level, resulting in new high-performance materials for use in transportation, computers, energy and communications.

3. Compact, long-lasting and highly portable energy sources, including fuel cells and batteries that can power electronic devices such as portable personal computers.

4. Digital high-definition television that will advance computer modeling and imaging.

5. Electronic miniaturization for personal use including interactive, pocket calculator-sized wireless data centers combining telephones, televisions and computers capable of storing and transmitting large amounts of information.

6. Cost-effective, adaptive "smart systems" that can integrate power, sensors and controls. These systems can control manufacturing processes from beginning to end.

7. Anti-aging products that rely on genetic information to slow the aging process. The products could include aging creams that really work or, more substantially, telomerase-affecting genetaceuticals that adjust our cellular clock speed.

8. Medical treatments that employ highly accurate sensors to locate diseased tissue and drug delivery systems that can target with great precision selected tissue and organs in the body.

9. Hybrid-fueled vehicles that are equipped to operate on a variety of fuels. These vehicles will select the appropriate fuel based on availability and driving conditions.

10. Education and entertainment including educational games and ultra-realistic simulations that will meet the sophisticated tastes of technology-literate consumers.

In summary, we believe that the major technology trends for the coming decade are

- Improved connectivity: The Internet is the beginning of a permanent and evolving communications infrastructure among people, information, products and the environment.

- The genetic diagnosis and treatment for aging and the prevention of all types of diseases.

- The dominance of nanotechnology: Nanotechnology is the science of manufacturing objects and structures with atomic precision, literally atom by atom. "Nano" is derived from the Greek word for dwarf. It is used as a prefix to indicate one billionth of a particular unit (i.e., one nanometer is one billionth of a meter). Ten hydrogen atoms laid end to end cover the distance of one nanometer. It is the first technology that has the potential for mastering all forms of matter and energy. It will have a tremendous impact when it is fully developed. Simple applications could involve the development of new and powerful materials. Just as nature is able to grow an oak tree from a seed, with nanotechnology it will be possible to manipulate objects on any scale. More advanced applications could involve such systems as nanocomputers, self-replication and biological and non-living nanodevices able to interact with their surroundings in very sophisticated ways. Nanotechnology could make it possible, at least theoretically, to assemble, disassemble and move atoms at will. Small entities could be created, atom by atom, that could accomplish many different tasks such as multiplying themselves, repairing themselves and creating larger structures. A push is under way to invent new devices that can be manufactured at almost no cost by treating atoms discretely, like computers treat bits of information. This could allow the automatic assembly of consumer goods without traditional labor, like a copier machine producing unlimited copies without a human having to retype the original information. Working smaller has led to new approaches for manipulating individual atoms in much the same way as proteins in a potato manipulate the atoms of soil, air and water to make copies of theselves. The combination of chemistry, biology, physics, electronics and engineering is ushering in the era of self-replicating, even self-designing, machinery and self-assembling consumer goods made from basic atoms.

# the virtual company

*I like the dreams of the future better than the history of the past*

—THOMAS JEFFERSON

WHAT IS A COMPANY? TRADITIONALLY, A COMPANY WAS COMPOSED of a factory or offices (physical spaces), various tools and machinery (everything from typewriters to turbines) and employees—the people who worked in the offices and factories and typed on the typewriters and operated the turbines.

In short, companies made things—tangible items you could see, touch, smell and even taste. And in making things, companies turned raw materials into refined products.

People mined coal, which was burned to smelt iron and power turbines, which produced electricity to power yet more machines. People used their hands and heads to make things. Reporters tapped out stories, which were printed and packaged as newspapers, which in turn were sold on street corners or delivered (often as a soggy pile of pulp) to doorsteps. Factory workers bent, welded and bolted steel and rubber into automobiles.

Accountants turned all this information into debits and credits, assets and liabilities. This language of the ledger book told the story of a company's tangible production, cost of goods and, ultimately, profits.

Additionally, companies traditionally were forged from the formidable wills—and wallets—of individuals. These capitalists, industrialists, robber barons and others often controlled more than one industry as a way to control production and repel competitors. They came up with the idea for a new product, then went out and created the system for making it in huge numbers.

Take automobile maker Henry Ford. Not only did Ford own the means (the factories) for producing his novel horseless conveyance. He also personally controlled many of the ancillary enterprises that supplied his company with materials for making his cars. He owned coal mines, which fueled furnaces in steel mills that he also owned. That steel was used in his factories to build Fords and later Lincolns and Mercurys.

You could say that Ford was his own supply and production chain. And Henry Ford wasn't the only one.

Such monopolistic chains of production no longer exist—they are far too expensive and inefficient to maintain. There isn't a major company today that doesn't depend on other companies for its existence or mutual co-existence.

Today, Ford Motor Company buys most of its parts from other companies, many of which are in other countries. Independent consultants are hired by Ford to tackle everything from balancing books to burnishing its public image. Ford even partners with other carmakers to produce autos—once a heretical notion. A failure by any one part of Ford's "network" of suppliers could significantly hamper the entire company, and would certainly ripple throughout the entire chain of Ford's business relationships with other companies.

This commercial interconnectedness applies to most businesses, which exist in symbiotic relationships with each other. To be sure, companies today could no more exist without a network of suppliers, subcontractors and strategic partners than they could live without the customers who buy their goods and services. This "eco-system" of dependent businesses forms and feeds the global economy.

## DAWN OF THE INFORMATION AGE

Along with this increasing shift from narrow machine age supply lines, today's companies find themselves making fewer "things." Many of the goods and services that companies produce today are considerably less tangible than the goods and services of years gone by.

Companies of the Industrial Revolution cranked out such marvels as steam engines and textile mills, and did so in fantastic volume through the wonder of then-fledgling mass production techniques. Today, companies find themselves making things of increasing complexity, but decreasing volume. This is the trend toward mass customization. Along with this change in what companies produce—more knowledge-intensive products and services—a change has emerged in the way organizations "think." The change in corporate "collective unconscious" is shown in Table 3-1.

## INFORMATION AS PRODUCT

Today, companies increasingly find themselves producing an entirely different type of product: knowledge.

If the mass production of goods was the way to wealth during the past two centuries, these same capital-intensive organizations and behemoth factories are now edging toward extinction. Knowledge is becoming the primary economic muscle of the information age. Today, nearly all the world's most innovative, successful and wealthy companies are those that wield knowledge effectively.

Microsoft, the world's richest software company, has *relatively* few physical assets (book value) that represent only a small portion

**Table 3-1. Organizational thinking—then and now**

| Item | Old Thinking<br>Industrial Company | New Thinking<br>Knowledge Company |
|---|---|---|
| People | Cost generators or resources | Revenue generators |
| Manager's power base | Relative level in organization's hierarchy | Relative level of knowledge and the ability to tolerate ambiguity |
| Power struggle | Physical laborers versus capitalists | Knowledge workers versus managers |
| Main task of management | Supervising subordinates | Supporting colleagues |
| Information | Control instrument | Tool for communication and development of knowledge |
| Production | Physical laborers processing physical resources to create tangible products | Knowledge workers converting information into tangible structures |
| Information flow | Via organizational hierarchy | Via collegial networks |
| Primary form of revenues | Tangible (money) | Equity & other intangibles (learning, new ideas, new customers, R&D) |
| Production bottlenecks | Financial capital & human skills | Technology, networks & intellectual capital |
| Manifestation of | Tangible products (hardware) production | Intangible structures i.e., new technology |
| Production flow | Machine-driven, sequential | Idea-driven, ordered-chaos |
| Effect of size | Economy of scale | Economy of depth/scale of the network |
| Customer relations | One way via markets | Interactive structures |
| Knowledge/technology | A tool or resource among others | The focus of business |
| Purpose of learning | Application of new tools | Creation of new assets |
| Stock market values | Driven by tangible assets | Driven by intellectual capital |

Modified from Svelby, 1997

of its staggering market capitalization. But its software products—the collective know-how of a group of computer programmers and the ability of the organization to bring it to market—are a runaway global success.

This ability to use knowledge effectively applies also to more traditional companies. Wal-Mart may be a retailing chain, but it uses the collective brainpower of its people to gain advantage over competitors. In fact, 90 percent of Wal-Mart's $200 billion market capitalization (8/99) is not attributable to book value. What is it attributable to? The short answer is intellectual capital or, more specifically, the market's valuation of Wal-Mart's ability to use its intellectual capital assets to generate profits. In today's knowledge-based economy, brains beat brawn almost all of the time.

Constructing a bridge or a house may seem like a pretty low-tech operation. Yet advances in construction and design technology—knowledge—are helping companies unearth fresh profits as they build better, longer lasting and safer structures.

Yesterday's raw materials were steel, oil, coal, products of the land and cash. Raw material in the information age is knowledge; its co-enzyme is information—the ability to solve problems, the ability to predict the future, the ability to convert human will into useful products and services.

When any age (or paradigm) gives way to another, old tools are shed as new ones are created. That is why a company's ability to acquire and develop knowledge, partially consolidated in the form of novel technologies, is vital to its survival in the information age.

## ORGANIZATIONAL THINKING—THEN AND NOW

**People:** *From cost centers to revenue centers*

In the past, employees were often regarded as a necessary inconvenience. Not liabilities, exactly, but not assets, either.

People were regarded as "cost generators." Sure, it was people who used their bodies to make a company's wares. But they cost money, first in the form of salaries and later in the form of salaries plus benefits. And people weren't terribly efficient—they got sick once in a while, and they were apt to balk at poor working conditions or low wages. With a touch of irony you might even say they were *ungrateful*.

Even with the development of machines of mass production, people were still needed to operate and maintain machines. Increasingly, companies needed fewer machine operators, but more people with specialized knowledge to keep the various systems humming along.

These days, in knowledge companies, people are no longer viewed strictly as cost generators. Sure, you still have to pay associates, but today many employees are viewed as potential revenue generators. Each employee contributes his or her knowledge; the result is a form of collective "brain" (structural intellectual capital) that solves problems that build a company's bottom line.

Sharing knowledge with others does not weaken the giver, because brainpower is not a depletable resource, except for the finiteness of an individual's life. Sharing knowledge with others, who in turn bring their own knowledge to the table, can boost knowledge exponentially. Everyone is smarter for it, including the company as a whole. When a knowledge worker solves a problem, he doesn't give up his knowledge, it actually grows stronger by virtue of the experience. You can think of this knowledge as an appreciating asset whose value is enhanced by experience and structural capital (information and knowledge networks and resources at the individual's disposal) and tempered by the willingness and ability to learn.

**Manager's power base:** *From relative level in the organization's hierarchy to relative level of knowledge and the ability to tolerate ambiguity*

Many companies aim to be meritocracies, rewarding the brainiest and most able employees with the best assignments and pay.

But tradition still pushes companies toward assigning workers according to school grades, tenure and other outmoded ladder-climbing criteria.

True, such yardsticks can work in selecting and promoting managers. But such means of measurement is outdated in the information age. This is especially true as the global economy becomes increasingly dependent on—and fueled by—brainpower instead of physical and mechanical brawn.

For companies to succeed in the information age, managers must tap into all available sources of information to mine for knowledge, which is a company's true asset. That means that companies must look at all employees as potential sources of knowledge. While traditional companies relied on knowledge to be passed down from management to staff, who performed tasks much as soldiers took orders, modern corporations need to tap into every available source of knowledge available.

This doesn't mean that managers need to put more suggestion boxes in break rooms. It means that companies need to tap into the collective brainpower of all their employees. It means that companies and their managers need to recognize how each employee's particular knowledge, not just tenure and perceived corporate loyalty, can help solve each new problem.

That is why selecting good managers is key. Companies should assign managers to various problems and jobs based on their own set of skills and knowledge rather than their hierarchical and bureaucratic pecking order.

Moreover, companies need managers who can swim in the shifting seas of the information age. Above all, managers must be able to tolerate the tremendous ambiguity of ideas and knowledge that they will face every day. In short, managers have to know how to manage knowledge—how to encourage its creation and development in others—rather than learn a particular mode of doing something and riding herd on underlings to do it in a prescribed way.

**Power struggle:** *From blue-collar versus white-collar workers to knowledge workers versus managers*

The traditional us-versus-them power struggle between employees and managers no doubt will continue in the information age. A global economy is no panacea for human conflict—it simply changes that conflict to friction between knowledge workers and this is why good managers—people with tolerance for ambiguity and change, and respect for the talents and knowledge of colleagues—are paramount to building a successful information age company.

**Main task of management:** *From supervising subordinates to supporting colleagues*

Managers aren't just corporate babysitters, passing out orders and meting out punishments for poor performance. Their success—and the company's success—hinges on how well knowledge workers share what they know and how they use this new knowledge. Managers are fast developing into knowledge facilitators, people who help assemble teams of knowledge workers to tackle various tasks. They're more true coaches and colleagues than backseat drivers.

**Information:** *From control instrument to tool for communication and development of knowledge*

For many companies in the past, information was considered a commodity to be made available on a need-to-know basis. That is, managers told employees only what they deemed necessary for them to do their jobs, and nothing more.

For a knowledge company to function well, information can't be used in a carrot-and-stick fashion. In fact, having employees who are out of the loop can be downright detrimental to a company whose fortunes rise and fall based on the quality of information available to all members.

**Production:** *From physical laborers processing physical resources to create tangible products to knowledge workers converting knowledge into tangible structures*

Workers once used their hands to make tangible goods. Natural resources (coal, iron and wood) were used to make physical structures and items of all shapes and sizes. But in the information age, workers rely on their brains to make their product: knowledge.

**Information flow:** *From organizational hierarchy to collegial networks*

You know the drill. Division heads get word from a CEO or other company manager that some new process or system should be implemented. These division heads meet with and relay some of this information (with bits about the larger corporate strategy expunged) to those who are next in the chain of corporate command: managers. In turn, these folks pass along an edited version of this information to employees.

This is fine if you're running an army, or a machine age organization. But it's a great way to bankrupt a knowledge company.

More open, collegial networks are a better way to manage information flow in today's knowledge-based companies. A company whose chief product is knowledge forged from the collective input of its employees needs to have a free exchange of information.

**Primary form of revenue:** *From tangibles (money) to equity and other intangibles (learning, new ideas, new customers, R&D, etc.)*

Companies once solely spun toasters and airplanes into profits. Now, many turn the raw materials of the collective brainpower of their employees into new ways to learn how to work better, create fresh products and attract and keep new, collaborative customers.

**Production bottlenecks:** From *financial capital and human skills to technology, networks and intellectual capital*

Today, the biggest encumbrances to corporate growth are often antiquated technology, attenuated networks of contacts and alliances and a dearth of the most important thing that makes a company hum in the information age: intellectual capital.

The sleekest, snazziest new techno-gadget won't necessarily boost earnings. But if competitors' new technology tools help them do better work more efficiently, you're at a significant disadvantage. Don't be a slave to technological novelty. But don't fail to recognize new tools that will help your company.

And in much the same way that collegial, collaborative communications are vital to a company generating knowledge, so it is that a company must work to grow the size and quality of its relationships with other corporate partners, suppliers and customers. Indeed, the need for a good network is true both in the microcosm of a company and the macrocosm of the global marketplace.

**Manifestation of production:** From *tangible products (hardware) to intangible structures (new technology)*

Workers once made products—tangible items you could smell, touch or taste. To make such goods required hands-on work.

*In the information age, workers use brainpower to fashion fresh knowledge. This newly minted knowledge is itself becoming the product, be it an improved process for managing a knowledge company or a fresh way to make cars safer. What once was intangible is now a tangible product* (Thomas Stewart, 1997)

The knowledge is no longer just a means to such tangible ends as cars or buildings or other products; now it *is* the product.

**Production flow:** *From machine-driven sequential to idea-driven ordered chaos*

Anyone who has seen a production line producing identical copies of a product one after another knows that consistency and quality are the name of the game. Each can of soup must be exactly the same as the last and totally free from defects.

While this is a fine and necessary structure for most machine age operations, it's a good way to kill an information age company, which by its nature thrives in an idea-driven world of ordered chaos (Figure 3–1).

Plodding, linear thinking won't fly in an era that increasingly rewards ability to make lightning-fast connections between ideas and concepts. Nimble, intuitive thinkers will thrive.

**Effect of size:** *From economy of scale to economy of depth and scale of the network*

In basketball, football, wrestling and some companies, size matters a lot. In knowledge companies, size is far less important—10,000 chimpanzees with typewriters can't write *War and Peace* any more than a hurricane hitting a junk yard can produce a Lear jet.

That is, the age of the mammoth company resembling a small country no longer works. Sure, big companies will continue to flourish. But the big, largely self-contained company of the machine age that designed, built and sold its products is dying, and is being replaced by a new corporate animal, one that relies on the depth and scale of its network of alliances (structural capital) for survival.

Knowledge companies are finding that it isn't how big your corporate headquarters is, but how many smart employees, vendors, corporate partners and alliances your corporate and personal network contains.

FIGURE 3-1
Machine age vs. information age

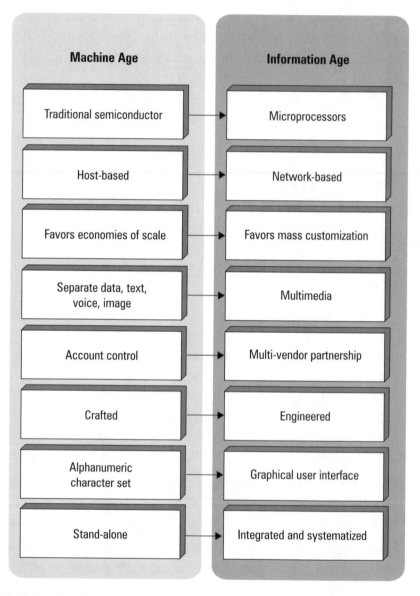

Modified after data from Ungson and Trudel, 1998.

**Customer relations:** *From one-way to interactive markets*

In the collective, bee hive-like atmosphere of the knowledge company, the line between customer and company is blurring.

Where once a company might have sold someone a car and wished them *happy motoring*, corporations now partner with clients in a symbiotic sharing of services. Clients partner with companies, collaborating on projects to foster better service and cement their relationships. It's much like inviting your best friend to dinner, and this friend happens to be a chef, who brings wine and helps cook. The food's better, and it's more fun.

**Knowledge/technology:** *From a resource to the focus of business*

Technology is no longer just a tool or a resource. Today, creating technology is itself a form of knowledge. And knowledge is a critical business asset.

**Purpose of learning:** *From application of new tools to creation of new assets*

In the past, a company would send an employee to a seminar or college course when more skills were needed to tackle a job. Often, after a three-day seminar, the employee was supposed to be an instant expert, ready to apply the weekend schooling on Monday morning. When problems cropped up, managers could always point to the pupil and say: "Well, we sent Bob to New York for training on that last month. He should be up on that stuff now."

These days, companies understand that learning creates fresh assets. That is, if knowledge is a company's product, learning is intellectual capital. Companies also are realizing that learning is a lifelong process rather than a quick fix. It's what business in the information age is all about. You don't graduate from college thinking, "Well, now I know everything I'll ever need to learn about that."

**Stock market values:** *From being driven by tangible assets to being driven by intellectual capital*

If Amazon.com's total physical assets were relatively small in August 1999 ($573 million), why was its market capitalization ($23.8 billion) so big, especially if its return on equity is −19.8 percent? Because the value of Amazon.com stock is driven not by how much real estate it owns, or how many books it has in inventory, but by an arguably more difficult-to-measure assest: intellectual capital.

## MACHINE AGE VS. INFORMATION AGE

Whether you call it the post-capitalist or post-industrial era or the information age, the emergence of a knowledge-based economy is no MBA school fad.

America's rise to world superpower status was built on mastery of machine age technology and business practices. But currently the wealth of nations and individuals increasingly depends on the ability to adapt to the ever-shifting conventions of the emerging knowledge economy. That means learning how to constantly adapt, rather than simply adopting a new set of conventions and rules.

The ability to adapt may be our most important survival skill. A story attributed to Peter Medawar, British zoologist, underscores the importance of being flexible. In the 1400s Spain could well have been considered the center of the modern world. She was the dominant naval power, was the center for trade, and in the process had amassed great wealth. Somewhat fittingly, the coat of arms of the Spanish Royal Family was a ship sailing between the Pillars of Hercules with the Latin motto inscribed beneath the icon, *Ne Plus Ultra* (No More Beyond). After Columbus rediscovered America, efficient court advisors quickly changed the motto to *Plus Ultra*. There is always more beyond. Be flexible.

# building university and government laboratory relationships—U2B™

4

*An invasion of armies can be resisted,*
*but not an idea whose time has come*
—Victor Hugo

B EFORE THE 1980s, UNIVERSITIES AND CORPORATIONS GENERALLY didn't mix very much. It wasn't that the two disliked one other. Rather, each group existed apart, constrained by law, culture and habit.

Universities did graduate people with MBAs and advanced science degrees, many of whom joined corporations, but the modern concept of technology transfer—that is, turning university research into commercial products and services—was at best an occasional, if somewhat unpredictable, outcome. There was, however, some overlap between industrial and academic goals during the years preceding the 1980s.

The Morrill Act of 1862 created "land grant colleges" to boost economic growth in the agricultural industry. Universities and U.S.-based companies both benefited from massive amounts of federal spending. Universities benefited because the government footed hefty bills for basic research—the sort of pure scientific exploration that few companies like to tackle alone because such work may never lead to profitable products or services. And corporations benefited in several ways. They profited from work done directly for the government, and they were able to use the government-funded research to "spin off" products and services to the private sector—a lucrative double dip. Still, the use of university-developed technology to fuel commercial industry was relatively unknown for decades.

The federal government funded research at hundreds of universities around the country. In turn, this research led to the creation of hundreds of potentially valuable patented technologies, most of which sat unused on shelves, the victims of both a lack of additional resources and limited links with industry for turning technology into commercial products and services. It was a bit like owning a gold mine before the invention of the pick ax.

Over time, however, Congress began to see the need for more effective technology transfer mechanisms, along with rules and procedures to govern these transfers. Nevertheless, less than a generation ago, the government was still unwilling and ill-equipped to license government-developed technology to the private sector. Red tape and overregulation based on cold war fears, as well as entrenched bureaucracies, exacerbated the government's reluctance to transfer technology to the commercial sector.

Moreover, there was a lingering perception among business people that the government had little incentive to seek practical uses for the inventions it owned beyond strategic defense initiatives. This began to change, however, after the winter of 1980, when Congress passed a bill aimed at bridging the chasm between government-funded university research and the private sector.

Sponsored by Senators Birch Bayh of Indiana and Robert Dole of Kansas, the bill had two purposes: (1) to allow universities, not-for-profit corporations and small businesses to patent and commercialize their federally funded inventions and (2) to let federal agencies grant exclusive licenses for their technology to provide more incentive to businesses.

Amendments to the Bayh-Dole Act created guidelines for tackling technology transfer and expanded the law's purview to include all federally funded contractors. What's more, the law made the first step in clarifying the roles and responsibilities among universities, the government and industry in performing technology transfers.

With the passage of the Bayh-Dole Act, Congress acknowledged that its research funding policies were dated, and perhaps even damaging to the economic future of the country. European governments had already been funding private research and development for some years, and the success of those efforts was noticed on this side of the Atlantic.

Congress also recognized that imagination and creativity, and especially scientific imagination and creativity, can and should be considered national resources. If scientists could help the government develop defense systems, Congress reasoned, why couldn't these same scientists and researchers help American companies develop superior products and services?

Key elements in the Bayh-Dole success story included:

◆ Universities (and other non-profit institutions), as well as small businesses, could elect to retain title to innovations developed under federally funded research programs.

◆ The use of inventions made possible through federal funding of universities was promoted and collaboration with commercial enterprises was encouraged.

◆ Universities were expected to file patents on inventions they elected to own.

◆ Universities were expected to give licensing preference to small businesses.

♦ The government retained a non-exclusive license to practice the patent throughout the world.

♦ The government retained "march-in" rights. That is, Uncle Sam could come knocking any time to see if these provisions were being observed.

The Bayh-Dole Act sparked an explosion of interest in technology transfer. Before 1980, fewer than 5 percent of the 28,000 patents held by federal agencies had been licensed (U.S. General Accounting Office, 1992).

The number of patents held by major U.S. universities has skyrocketed too—from 230 patents issued in 1976 to more than 2,500 in 1996. More than 200 universities around the nation are now involved in technology transfer—eight times as many as before the Bayh-Dole Act.

Moreover, it is estimated that these technology transfer efforts boost the economy by more than $24.8 billion and 212,500 jobs annually. Such technology transfer efforts are spawning new businesses and helping to further expand existing markets. Plus, new types of industries are being invented.

Still in its infancy, technology transfer appears poised to bloom in coming years. A 120 percent increase in the number of U.S. patent applications—along with a 68 percent surge in licenses between fiscal year 1991 through 1995—point to the booming growth of technology transfer efforts. In particular, commercialization of university-developed technologies under the Bayh-Dole Act spawned the biotechnology industry and led to significant commercial advances in other technology-intensive industries.

In 1997, university R&D was responsible for the development of 44 percent of all new U.S. drugs and 37 percent of all new pharmaceutical processes. Information and communications industries have also benefited, with 28 percent of their products and 27 percent of their processes being spawned at least in part by academic research in 1997. Even the metals industry got a boost from university research, which was responsible for 22 percent of the metals industry's products and 21 percent of its processes.

A 1998 survey of universities by the Association of University Technology Managers, Inc. (AUTM)—a nonprofit group created to help university intellectual property managers—found that the Bayh-Dole Act's economic effects are "extensive."

Among the report's findings:

◆ The number of U.S. patents issued to universities increased to 3,224 in 1998—an increase of 22 percent over 2,645 patents in 1997.

◆ Universities reported having 17,088 active licenses or options —an increase of nearly 11 percent over the previous year.

◆ During this period, licensing income increased from $611 million to $725 million.

◆ 11,784 invention disclosures were reported.

In 1998, U.S. universities received $24.4 billion to support research. Of this, $15.3 billion was provided in direct federal funds for science and engineering research and development, not to mention the additional access to billions more through affiliations with other nonprofit organizations.

Table 4-1 shows specific federal agency contributions to U.S. university research funding in 1999.

## VAGUE BY DESIGN?

The Bayh-Dole Act may seem like a heavy-duty how-to manual for technology transfer deals (as well as a rulebook for policing these

**Table 4-1. U.S. funding to colleges and universities for research in 1999**

| Agency | Amount of Awards (In Thousands) | Percent of Total |
|---|---|---|
| Health and Human Services | $7,446,857 | 60.90 |
| National Science Foundation | $2,149,853 | 17.58 |
| Department of Defense | $1,082,658 | 8.85 |
| NASA | $596,090 | 4.87 |
| Department of Energy | $552,825 | 4.52 |
| Department of Agriculture | $399,709 | 3.27 |
| Total (All Agencies Listed) | $12,227,992 | 100.00 |

Source: National Science Foundation, 1999.

policies), but it is really more of an outline. Indeed, no one agency or organization oversees Bayh-Dole policies. Rather, administration of the Act is decentralized, with federal agencies acting as facilitators more than police.

Universities tackle administrative duties; federal agencies collect and process paperwork. This happens more by fact than fiat. Each funding agency administers the law as it applies to its own grants and contracts. If a problem comes up, say, a conflict between an inventor and his or her university, the funding agency may choose to mediate, but more likely the problem will be solved at the research institution level. Day-to-day operations of the Act have largely been left up to the universities. Indeed, even a comprehensive national study of how many U.S. universities have set up programs to push technology transfer efforts has yet to be done. The U.S. Government Accounting Office studied 10 universities that received among the largest federal research and development and licensing income in

1996 and found that all had established programs specifically aimed at implementing Bayh-Dole Act policies.

## CURRENT EFFORTS

So, how do universities orchestrate technology transfer? Bayh-Dole says plenty about what universities, the government and industry should be doing, but it says curiously little about just how to do it. It's a like telling folks on either side of a river that they need to visit each other, without offering to show them how to build a bridge. Without a blueprint for how to use Bayh-Dole, universities typically use one of the following methods for identifying inventions to be licensed and for offering them to prospective clients.

◆ *Centralized licensing*: All activities are funneled through a single university office. (MIT uses such centralized licensing.)

◆ *Decentralized licensing*: Licensing activities are performed by several offices and associated with a university's various schools and departments. (Johns Hopkins has a licensing office for its medical school, one for its Applied Physics Laboratory and one for the rest of the university.)

◆ *Foundation*: Licensing is done by an independent foundation set up specifically for this purpose. (State universities such as the University of Wisconsin-Madison favor this arrangement.)

◆ *Contractor*: A contractor is hired to handle licensing. (Michigan State uses a Tucson, Ariz.-based corporation.)

*Source*: United States Government Accounting Office, 1992.

## TRADITIONAL TACTICS

Typically, university patent/licensing directors sift through mountains of disclosures, patent applications and licenses, hoping to discover technological gems that can be presented to companies—a painstaking and imprecise operation at best.

Smart, techno-savvy types though they often are, these arbiters of what will and won't be presented to companies often work solo or have very small staffs—not the most efficient way to examine several hundred scientific discoveries each year. So naturally they hunt for big game; scientific discoveries with obvious, ready-made commercial applications.

They also tend to target large, established companies as potential customers, such as General Electric, Bayer Pharmaceuticals and others with enough money and marketing muscle to increase the odds of turning science into a profitable product or service.

This seems sensible, but it overlooks the fact the corporate giants have large research and development operations of their own. Because these companies spend billions each year on inventing and nurturing hundreds of their own R&D projects, they are less apt to give undeveloped technologies from outside much of a look. This is the paradox of courting the big company, the "safe bet" customer. It's the lotto mentality. All you need is a dollar (the price to send an unsolicited letter to a large company) and a dream. And sure there are winners. Every lottery has them. But this is no way to build an efficient market between buyers and sellers.

Imagine an in-house scientist telling his boss, "You know the work you guys have spent so much money on? Well, this guy outside our company has come up with far superior science. I say we dump our work, cut our loses and license his patent."

Moreover, big companies often would rather wait until a technology is "proven" to be marketable before it opens its wallet. By then, a breakthrough invention may have been surpassed by competing sciences or may simply have been mothballed for lack of interest and investment. One large pharmaceutical company indicated that it was easier for it to invest $200 million in a product that had passed the first level of FDA approval than to invest $2 million in a powerful new idea with great potential.

Plus, what university has the time and money to present an invention to everybody, especially to thousands of small- and medium-sized companies? Marketing new science takes time and plenty of shoe leather. That's why university licensing directors concentrate efforts on what they see as "sure things."

What's more, smaller companies may lack the cash to buy technology. And even if they can pay the tab, they may not have enough money or scientific talent to develop the technology into a product or service, or to market it successfully.

Of course, deals do get done—a combination of habit and commercial conventions. Most technology transfer deals work this way:

The university that owns the patented technology (the inventor normally has a royalty sharing agreement with the university) negotiates the sale of a license to a company. Typically, the agreement uses a combination of lump sum fees paid to the university, along with a percentage of royalties earned from the sale of a product or service. Rates and fees vary, depending on the value of the technology, the marketplace demand and the skill of negotiators on each side.

Typically, universities receive between 1 and 10 percent of a company's gross sales on a licensed technology, with 2 to 5 percent being more typical. Inventors usually earn between 10 and 50 percent of what their university employers receive after legal expenses are deducted.

On a product that produces $100 million in sales, a university may earn $5,000,000 over a number of years. From this, the inventor may receive $500,000 to $2,500,000. In addition, as is often the case, royalties may not appear for a decade or more, until a company finally refines, produces and markets its licensed technology. Long term, the major value in a technology license is determined by the collective royalty stream. Traditional technology transfer companies normally "charge" or receive 50 percent or more of the long-term royalty stream in consideration for their services. With this in mind, the above payment distribution scenario would be cut in half. Indeed, with most traditional technology transfer approaches, universities receive but a small portion of a company's sales from licensed products, and inventors get even less.

Under the current model, it's a little like a person who's having trouble selling his house being approached by a house broker, who says, "Assign to me the deed and I'll give you 50 cents on every dollar I earn from the sale of the house."

No sooner would the broker say this than he'd find the door slammed in his face. Yet this is essentially the kind of deal most universities cut when they work with traditional technology transfer companies. If you're willing to tolerate some ambiguity, you may feel that real estate and intellectual properties are not that different conceptually. They are both properties. They have long-term and present values. They both can be improved, leased, licensed, sublicensed (sublet), transferred, escrowed, bartered and sold. When unproductive, they cost their owners money in the form of taxes, maintenance and management fees.

Nevertheless, under the current technology transfer systems, universities and inventors sometimes earn significant dollars, but as mentioned before, the system largely resembles a lottery (most players do not win—most technologies do not get licensed) and when they do, the universities receive only half of their long-term value in the property as a result of technology transfer company royalty sharing agreements.

Some technologies are snatched up because they are obviously useful in a big way. Others languish for lack of marketing or because they simply don't appear immediately useful. Or maybe the technology's usefulness would be clearly apparent if it could be developed just a bit more.

Consider this: The top 160 U.S. universities, medical centers and research institutes spent about $21 billion on research and development in 1996 and produced about 10,000 disclosures and 2,500 patents—or about 200 new technologies each week. Those top universities produced more than 90 percent of all university-based inventions. Yet more than 70 percent of these patented technologies go unlicensed. Moreover, with a couple of notable exceptions, all universities spent far more money creating technologies than they ever earned from licensing. Actually, the average return on equity (ROE) for all universities in the AUTM study was about 1.5 percent before expenses. As a group they lost a large sum of money, with an even larger opportunity cost.

The bottom line is that universities produce vast amounts of technology that never gets to market. But what if an alternative arrangement existed, one that would satisfy both companies (especially small ones eager for technologies) and universities, and one that would offer greater rewards and incentives for further research to both universities and inventors? And what if universities no longer had to spend a dime on marketing their technologies to technology corporations?

## A NEW PATH

The new model described in this book allows universities to receive 100 percent of royalties earned on inventions. Plus, the model ensures faster, and therefore potentially more profitable, execution of licensing agreements and gives universities a way to market their technologies to a vastly wider market at no cost to them. Moreover, this novel model of technology transfer allows universities to focus on their strengths—basic scientific research.

Companies also win. With this model, small companies can boost core technology holdings and acquire powerful new products without utilizing their limited cash. Furthermore, companies can pursue high-quality R&D work and foster closer ties with universities and government labs (the idea factories) that have a vested, focused and long-term interest in their technology franchise.

Unlike previous methods of technology transfer, this approach is the first to offer a mechanism for financing the acquisition and development of technologies. What's more, this new tool for financing technology transfer could be used to turn more unused technologies developed at universities into innovative products and services— helping companies, universities and, ultimately, the public.

We'll take a closer look at this model in Chapter 8. But first, let's see how current university research is funded.

## WHO PAYS FOR RESEARCH

If companies fund the development of scientific discovery, who pays for the original science?

Increasingly, universities are coming under fire about the value of various types of research, of which there are essentially two: basic and applied.

**Basic** research, as its name suggests, concerns inquiry into the fundamental questions of nature—what makes some cells more or less susceptible to cancers or whether anti-matter "exists." In other words, the big questions.

**Applied** research employs scientific methods to solve a specific puzzle. It means using tools such as knowledge about why some cells resist infection to come up with practical cures. You could say that applied research uses big question answers to answer smaller queries.

Debate is ongoing about whether companies should shoulder more of the costs of applied research, as well as arguments about the value to the nation of government funding of basic research.

Some argue that government funding of research at universities and non-profit facilities is inherently unfair. They contend that this is tantamount to giving away the nation's most precious commodity—knowledge—for free.

Moreover, critics claim that when the government helps pay for applied research, or when companies help fund R & D efforts at universities, these actions distort the priorities of science. These same critics claim injecting a profit motive distorts the mission of university scientific research.

No doubt corporations may enrich themselves with university research. Yet who could argue that the government's help in paying for a company's research into finding a cure for cancer or AIDS doesn't have a public health benefit that outweighs concerns about distortion of the goals of basic or applied research.

Sure, a company may earn a significant amount from a new cancer-fighting drug whose development was funded in part or in whole by the government. But don't benefits to life and health outweigh concerns about who pays for what? Indeed, doesn't it make the world a richer, safer, better place to live? Few university innovations would have a significant impact in the marketplace without private sector financing.

Keeping universities and private sector corporations separate from each other may be better for both (here is where the bridge analogy is particularly useful—separate yet connected), as well as for us all. Even if one can't live without the other, that doesn't mean living under one roof is a good idea. To be sure, in our experience, university research scientists don't normally thrive in a corporate environment.

Being married to one corporation would also stifle a university's creativity and ability to share knowledge in an open, free fashion—often the very antithesis of corporate competitiveness. Universities do best when they are able to make connections when it serves their interests. This is also true of corporations, who can tap into a university's knowledge base when needed, but don't have to fund all of the research projects.

It seems apparent, too, that using public funds to foster advancements in science and technology boosts the entire nation's competitive environment, while it helps solve today's applied and theoretical science problems. Public funds also are a great source of new technologies to help grow the business segment responsible for the greatest number of new jobs—small companies.

# intellectual property

5

*The more original a discovery, the more*
*obvious it seems afterwards*

—ARTHUR KOESTLER

L ET'S SAY YOU ARE A SCIENTIST AT A UNIVERSITY, OR PERHAPS A
corporate executive whose in-house lab has created a great
new technology that will almost certainly revolutionize the way
people communicate. With visions of astronomical profits—to be
followed by world peace—dancing in your head, you begin planning
a conquest of competitors.

But before you market your new technology, or so much as whis-
per a boastful word at the next cocktail party, you had better make
sure you do something to prevent your competitors from beating
you to the punch. In other words, patent it.

But wait, is patenting the idea really the right thing to do?

To settle the question, let's look at what comprises "intellectual
property"—creations of human intellect that are protected by law—
and how companies can best safeguard it.

There are several types of intellectual property: patents, copyrights, trademarks and trade secrets. Copyrights protect literary and
artistic works (books, music, movies and the like); trademarks, also
known as "brands," are words, designs or other symbols that

identify and distinguish products and services. Both copyrights and trademarks are registered with the U.S. Patent and Trademark Office.

Then there are patents.

## WHAT A PATENT IS AND ISN'T

The word "patent" has meant different things over the centuries, but today it is most commonly used to refer to an official document that conveys specific rights granted by the U.S. government to a person or group (the "patentee"). These rights are called "patent rights."

Originally conceived as a way to spur invention and investment, patent laws, along with copyright laws (the literary equivalent of a patent), were among the earliest laws created in the United States (1790). Patents and copyrights are authorized in the U.S. Constitution to "promote the progress of science and useful arts." Patent rights give incentives to inventors and their employers to create new technology and to invest in commercializing technology. Policy makers have generally agreed that the U.S. tradition of strong patent laws has made a significant contribution to positioning the United States as one of the world's leading technology developers.

Basically, a patent is a legal monopoly. With it, you are given the right to exclude others from making, using or selling your invention throughout a particular territory. The owner of a patent may be the inventor, but he or she need not be. The inventor's employer, or someone who has acquired the rights from the inventor or the employer, can be the owner. Indeed, universities often license their technologies to companies, which use them to make products and services in return for payment and/or royalties.

To be eligible for a patent, an invention must be "new," and must be sufficiently different that it is not "obvious" to a person skilled in the art. This doesn't mean an invention has to be an improvement, only that it be distinguishable. Every kind of applied technology can be patented, but scientific principles and naturally occurring materials are not patentable.

Curiously, patents do not guarantee the right of the patentee to make, use or sell the invention. Indeed, in some instances, a patentee cannot make, use or sell the invention without a license from yet another patentee. Sometimes, patenting an invention actually violates the prior patent rights of a patent owned by someone else. This is known as "infringement."

Picture this scenario:

*You hold the dominating patent on a motor-powered vehicle—a preemptive patent covering every type of motor-driven vehicle, including cars, trucks, etc. Another person comes along a couple years later and obtains a patent on an internal combustion engine.*

In such a situation, the second patentee, the one who invented an internal combustion engine that would make a car run, cannot make or sell such a vehicle unless he gets permission from the dominant patentee.

Conversely, because the second patentee has the only practical means of powering the vehicle, the patentee of the vehicle itself cannot make or sell a motor-driven vehicle in the patent's territory with this internal combustion engine without infringing on the engine patentee.

So each patent holder needs the other to make a car. And because such cruxes are common, companies and inventors often must collaborate in order to produce anything at all. For example, hardly a single complex product produced by a major company would be possible without an elaborate network of collaborative cross-licensing deals with other inventors and manufacturers.

This is becoming increasingly true because systems of all kinds are too complex to exist without borrowing (or treading on) the intellectual property of others.

A patent is a powerful tool, but it may not be the best way to secure the competitive advantage that companies often live or die by. Sure, a patent gives an inventor a monopoly for 20 years. But, in the patent application, you also disclose to all competitors just how the proprietary gizmo or process works. In short, it's no longer a secret.

(You also can obtain patent rights in most foreign countries, but the laws can vary wildly from ours.) While you can always take violators to court, the costs can be stratospheric. Court tussles have scotched the launch of many promising products or services.

And some patents take precious time to get—on average about two years. You may shell out thousands of dollars to patent a particular invention, only to find that by the time you've wound your way through the process and are granted the patent, several years have gone by—an Ice Age in a marketplace that measures cycles of innovation in what seems like nanoseconds.

This is the paradox of patents. You seek a patent to lock out would-be competitors, but by the time you get one on a genetically engineered rose, the blush may already be gone.

The longer it takes to bring an invention from concept to consumers, the greater the chance that enthusiasm among researchers and developers will wane. It's this creative energy that propels the idea along, that carries a new product or service to the marketplace. Inspiration fuels innovation. The intellectual property of a research team is only one aspect of its intellectual capital. The most important aspects are the ability and willingness of the team to work together to overcome the inevitable obstacles, solve problems and figure out how to make the technology work in a practical and consistent manner.

For these reasons, in certain instances inventors may be better off keeping mum about their inventions (this is a difficult call to make, and expert advice should be sought to carefully weigh the benefits of secrecy over enabling disclosure) as they race to bring fully developed products and services to the marketplace ahead of competitors.

## NUTS AND BOLTS OF GETTING A PATENT

### What can be patented?

An invention can be patented only if it fits within one of the following classifications or conditions:

◆ It must involve an innovative *process*, which can mean an art, method or mode of operation and includes the novel use of a known process, machine, manufacture or composition of matter. A patentable process can involve the use of old steps that are used in a new way to create a fresh way of making or doing something.

◆ It must be some sort of new *machine* (either hand-operated or automated), which means an engine or apparatus that does something or produces some effect when activated. A patentable machine may crank out loaves of bread or mathematical computations.

◆ It must involve a novel *composition of matter,* which pertains to mixtures of ingredients, chemicals or physical elements that produce a defined effect. A newfangled cocktail of chemicals used as a medicine is an example.

◆ It must involve a new *manufacture,* which is a way of defining any number of inventions that don't fall under the other statutory categories. This can include such difficult-to-define products as building structures and designs, sound recordings and even genetically engineered organisms (Muir, 1997). Other examples might include software, although some types of software are unpatentable, such as those that involve the invention of a mathematical algorithm.

As new technologies are created, questions about what can and can't be patented arise almost daily. In particular, debate is growing over the extent to which genetically engineered inventions are patentable. For example, can the manipulation of genes to create a clone of a human being be patented?

## Defusing patent time bombs

While most of the work in obtaining a patent involves filing paperwork and crafting precise descriptive terminology, pitfalls exist that can wreck even the most scrupulously built patent application. One of the potentially most pernicious of these involves something called "prior art."

Say you're an inventor who has created a nifty new invention. You decide that getting a patent is prudent. So you begin jumping through the application hoops: filing papers, hiring an intellectual property lawyer to make all the necessary disclosures, etc.

As luck would have it, a conference is coming up that would be a great place to describe the scientific merits of your invention. Because you're well within the one-year "grace period" by which you must file a U.S. patent application after publicly disclosing the nature of your invention, you feel safe in talking about it.

Think again, because, while you are perfectly correct in thinking you can publish to your heart's content on the airwaves and the Internet about the great things your invention does, there is nothing to bar folks outside the United States from snatching your invention up and using it themselves. In effect, you may have just given your invention away free to the very global market you hoped to tap.

# leveraging
# intellectual capital

6

*Measure wealth not by the things you have,*
*but by the things you have for which*
*you would not take money*

— ANONYMOUS

L IKE MOST INVESTORS, YOU HAVE NO DOUBT REGRETTED NOT BUYING some profitable stock when it was trading in the single digits. Perhaps you have indulged in a little masochistic math, as well. If only you had bought a measly $1,000 worth of Cisco stock when shares were really cheap, you would be on your own private island today. If only you had taken a few minutes to actually read that Cisco prospectus, you might have seen the clues pointing to how the high-tech newcomer would someday become one of the richest companies on this planet.

Of course, the fact is that you probably wouldn't have anted up a single buck on a startup with such seemingly anemic assets. Why? Because chances are there was nothing in that prospectus that would have said anything about Cisco's most important holding. Traditional accounting methods may balance the books, but they reveal nothing about a company's most valuable asset: intellectual capital.

Intellectual capital is the collective brainpower and experience that can be used to create wealth for a company. It is the combination of individual knowledge and experience interacting with internal and central structural capital (Figure 6-1). For most technology companies, the chief assets are clever employees. Working together, they add something to a product you can't see, smell, touch or taste—knowledge.

This doesn't mean that the fruits of intellectual capital of all companies are as "intangible" as Cisco's; a company with an abundance of intellectual capital may make truck parts or canned hams. But companies that recognize the value of intellectual capital almost always share a common approach—they use the collective know-how of their employees, vendors and customers to create superior

**FIGURE 6-1**
Book value vs. intellectual capital

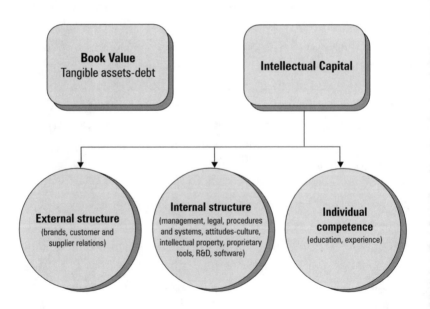

Modified after Svelby, 1997.

(and usually more profitable) products. They understand that brains and inspiration equal gold in the form of profits—*virtual alchemy*.

Today, corporations increasingly understand that they must leverage their intellectual capital to outmaneuver competitors in the marketplace. And they understand that intellectual capital is more than the sum of employee capabilities. Intellectual capital is a company's entire network of contacts, from old college buddies to subcontractors to customers to family members—in short, everyone who can partner with them in a symbiotic sharing of services.

But intellectual capital isn't simply what goes down the elevator at 6 o'clock every workday. And neither is it simply a chorus of employees plus vendor staff plus clients. Rather, it is the knowledge, technology, energy, experience and degree of positive collective purpose among all these people, leavened with a high tolerance for the ambiguity of new ideas, as well as the intuitive firing of neurons. All of that combined adds up to intellectual capital.

Creating and using intellectual capital means being able to work fast together. This doesn't mean you and your employees need to aspire to some sort of perpetual caffeine-fueled frenzy. Rather, it means competing effectively in a knowledge economy—thinking, acting and changing direction as quickly as ideas occur, in the context of sensible leadership.

That's because ideas and intellectual capital can have very short shelf lives. A super new idea worth millions today may be stale bread as early as next year. As we move into the information age, this premium on hair-trigger reflexes and compacted production cycles will become more critical for success and survival.

## THE KNOWLEDGE ECONOMY REVOLUTION

Okay, so maybe you missed out on a plum opportunity to buy Dell stock cheap because you couldn't read between the lines of traditional financial reports to perceive hidden values. You weren't the only one. Remember that in recent years scores of economists and business writers saw downsizing and layoffs as harbingers of fiscal doom and gloom, yet American businesses are leading the charge into the information age.

And high-tech companies aren't the only ones that are out in front. Curiously, many of the big, traditional manufacturing outfits are pioneering new ways to harness knowledge for profit. Indeed, the majority of the market value even for companies like home-products behemoth Procter & Gamble Co. and restaurant giant McDonald's comes from intellectual capital (97 percent and 84 percent, respectively).

You could say that trying to create intellectual capital is a lot like trying to separate super students and slackers from a pile of college admission applications. Safe, traditional bets will be on the straight-A kid with the meter-long list of extra-curriculars. But an intuitive hunch on a candidate with bifurcated SATs and an oddball array of hobbies could mean the difference between an alumnus who is a bright corporate grinder or an unconventional billionaire. The difference between success and failure might be better measured in relation to the desire to succeed rather than SAT scores alone.

Just as students bring different talents to a college campus, so do the various parts of a business and its wider knowledge network contribute to a company's intellectual capital. So, what is this mysterious dynamic that turns people and things into intellectual capital?

## HUMAN CAPITAL

Employees aren't the fodder they once were—cheap, replaceable cost generators who toted, fetched and turned raw materials into cars and other tangible goods. Today, people are quite literally the flesh and bones of a company—or, rather, the brains. Their collective know-how, experience and even institutional memory are what make a company possible.

Plenty of CEOs pay lip service to the notion that people are a company's greatest assets. But this isn't just another hokey slogan from a company handbook. If, to paraphrase Yeats, "human beings are the wellspring of innovation," then human capital is the most important part of a business built on knowledge.

Indeed, a company isn't a place where people go to pull levers and drill holes; it's a kind of conceptual rallying point for encouraging dynamic thinking. Human brains, not brawn, are now the engines that drive the economy; the importance of attracting and nurturing the talents of this human capital is paramount. Successful knowledge companies constantly work to harness, and not choke, the

## FIGURE 6-2
Components of intellectual capital

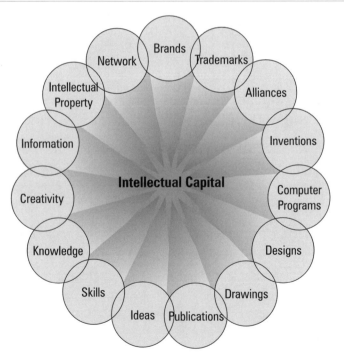

talents of employees and managers toward collective learning and improvement.

> If a company isn't downright tireless about tapping the creativity and innovation of its employees, it will die. And managers must be catalysts—they must unleash creativity rather than corral it. The need for inspired leadership is dramatically enhanced in the knowledge enterprise.

Curiously, while humans are a company's most valuable assets, they are assets that cannot be owned, per se. Cash and machines may be owned; people cannot.

Careers are built differently today than they were just a few short years ago. Employees once boasted of being IBM people, of bleeding blue. Today, companies come and go in the blink of an eye. Having a career no longer means being tethered to a single company. Successful companies collaborate with other companies and workers when it makes sense for each. Successful knowledge workers increasingly associate themselves with their profession. Fealty to one's field has replaced loyalty to a company.

## STRUCTURAL CAPITAL

You could say that structural capital is what hangs around the office once human capital has gone home for the night. Structural capital describes many things of varying degrees of physical solidity, all produced in some fashion by human capital.

First, it includes the legal framework for ownership: inventions and their patents and publications and motifs protected by copyrights and trademarks. It also includes the culture and reputation of a business—its brand, if you will. The signature way a company does business, its processes and systems for getting things done, also are part of its structural capital.

A company's structures include the way it deals with clients, partners and even competitors. In turn, all of these factors contribute to structural capital. While codified, structural capital is in many ways as mutable as the human capital that made it.

## CUSTOMER CAPITAL

Arguably the least tangible type of capital, customer capital, is basically a way of describing something that has kept companies and clients together for eons: goodwill. A company may be full of talented employees, cutting edge patents and universal name recognition, but without a strong and loyal customer base, it's nowhere.

There are practical reasons for wanting to nurture long-term relationships with clients, and making more money ranks pretty high on the list. Companies spend lots of money courting clients; keeping them typically costs less. Plus, as client needs grow, so does a company's potential income. As you learn new skills from your clients, you can use that knowledge to further relationships with other clients or snag new ones. A happy client will tout your talents to other potential customers. An unhappy client will tout your failings, only somewhat more efficiently.

For example, let's say your steel-making company is helping design a stronger, lighter-weight truck frame for a vehicle manufacturer. Chances are, both companies will gain plenty; your company will gather knowledge about improved methods for making and molding metals, while your client will learn novel vehicle construction processes. In this way, you both learn things you can apply to other clients, be they makers of steel or cars, synthetics or canoes.

The more you learn about the business of others, the better you will be in courting and keeping clients of all types. You become smarter, more versatile and potentially more profitable.

Goodwill, or customer capital, is the glue that keeps a network together and growing. The strength and scope of a company's network overtakes traditional hard assets in importance as we move into a knowledge-based economy—and customer capital will become a precious commodity.

Intellectual capital is the forging together of knowledge, ideas, creativity, experience and skills in a way that enables an enterprise to use these assets to make money. Some examples are developing new processes, obtaining patents, acquiring new and unique data, disseminating information to create brand awareness or strengthen a distribution channel and developing better processes—new ways to enhance productivity (Figure 6-3).

**FIGURE 6-3**
Creation of intellectual capital

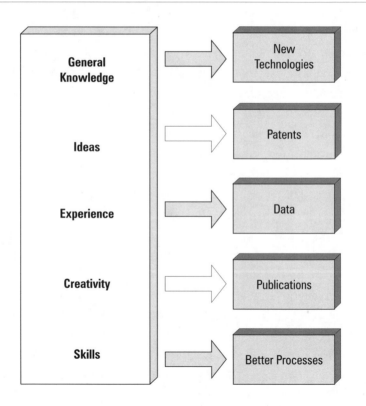

## MEASURING INTELLECTUAL CAPITAL

How do you measure how much (if any) intellectual capital your company has? Intellectual capital doesn't exist on traditional balance books. No decoder ring or X-ray glasses will pierce the secrets of these documents.

You will need to examine a combination of orthodox account ledgers and other, tougher-to-measure factors to tease out intellectual capital figures. Even then, you'll be making an educated guess. Measuring the intellectual capital of a successful company brimming with it is a bit like trying to gaze on an electron. No matter how much you increase the power of the microscope, you will only see the evidence of its existence. Karl Erik Svelby (1997) has assembled an approach to monitoring intangible assets that may be helpful (Table 6-1).

A corporation that successfully employs and expands its intellectual capital reaps profits. Another corporation with numerically equal measures of intellectual capital may earn a lot less. The fruits of inspiration are obvious; the quality of inspiration isn't.

For the purposes of this book, which presents corporations with a novel way of building fruitful relationships with universities to find and develop new technologies, we'll leave the metrics of intellectual capital to others. Still, the evidence of intellectual capital is glaringly evident when you look at company bottom lines.

Intellectual capital can be described as the difference between your market capitalization (the total monetary value of your company's stock right now) and your book value (what's left over if you sell all your company's assets).

For example, at the end of August 1999, Procter & Gamble Co.'s stock was worth more than $126 billion. Its total physical assets (buildings, plants, equipment, etc.) are worth a bit more than $3 billion. That means a staggering 97 percent of P&G's value—$123 billion—is in intellectual capital or what the market believes is the ability of P&G to harness its experience, brands, distribution system and know-how to get the job done.

**Table 6-1. Measuring intangible assets—implementing systems**

| Action to Take | Relation to Intangible Asset | Way to Monitor |
|---|---|---|
| **MANAGE COMPETENCE** | **BENEFIT** | **INDICATOR TO USE** |
| Carefully recruit bright people who are receptive to training | Provides inflow of new fresh competence and energy; strengthens culture | Rookie ratio, divided into university-educated rookies and other rookies |
| Improve the education level among all employees. | Increases flexibility and learning rate | Education levels |
| Offer careers that are up-or-out; no plateaus allowed. | Individuals encouraged to maintain steep learning curve or leave; creates turnover | Individual grading; average marks |
| Make competence maps. | Identifies competencies | Number of people in each category, of years in profession |
| Use juniors as assistants. | Enables tradition of tacit knowledge; reduces administration | Proportion of junior time spent for competence-enhancing customers |
| Keep people from leaving by creating loyalty. | Competence is not lost to competition | Attitude surveys; staff turnover |
| Build close personal relations with a few selected clients. | Creates inflow of knowledge customers | Proportion of competence-enhancing |
| Let young learn from old in master apprentice relationships. | Enables tradition of tacit knowledge | Attitude surveys |
| **MANAGE THE INTERNAL STRUCTURE** | **BENEFIT** | **INDICATOR TO USE** |
| Go for large assignments that allow teams. | Teams allow tradition of tacit knowledge among members | Proportion of organization-improving customers |
| Develop own concepts and methods. | Creates new knowledge, R&D | Time devoted to R&D |
| Publicize the concepts in books and seminars. | Influences the mindsets of potential customers; creates standards | Time devoted to such activities |
| Create information-sharing system. | Supports knowledge combination | Investment level in IT |
| Charge for teams, not for individuals. | Reduces internal competition; improves tradition | Proportion of team billing |
| Proactively manage age structure. | Reduces risk for plateaus, keeps balance between dynamic and static forces | Median age; staff turnover |

*continued*

**Table 6-1. Measuring intangible assets—implementing systems** *(continued)*

| | BENEFIT | INDICATOR TO USE |
|---|---|---|
| Build close personal relations with customers that provide R&D projects or large projects. | Improves internal structure and enables learning | Proportion of organization-improving customers |
| Encourage piggybacking in all departments. | Enables tradition of tacit knowledge | Proportion of junior time spent for competence-enhancing customers |
| Organize departments as open-space offices. | Enables tradition of tacit knowledge | Attitude surveys |
| Communicate mission for business. | Gives focus and purpose of knowledge creation | Attitude surveys |
| **MANAGE THE EXTERNAL STRUCTURE** | **BENEFIT** | **INDICATOR TO USE** |
| Focus management information on customers rather than markets or products. | Knowledge flows through relations, not through markets | Proportion of image-enhancing customers |
| Build image as "USA's most competent in your industry segment" by giving seminars, etc. | Reduces marketing costs | Number of seminars held; customer surveys |
| Select customers that contribute to intangible assets or profit; cut out the rest. | Concentrates efforts to most valuable customers; improves inflow of knowledge | Categorize customers, compute profitability, sales per customer |
| Build teams with customer chemistry in mind. | Improves success rate and inflow of knowledge | Win/loss index; satisfied customers index |
| Nurture image as an important asset. | Reduces marketing costs | Money spent, time used; satisfied customers index |
| Treat former employees as honored alumni. | Retains relationships that enhance instead of damage image; can also lead to new customer relationships | Alumni surveys |

(Svelby, 1997)

While this is extraordinary, such enormous market-book ratios are increasingly common for today's richest corporations. Every day, corporations of all kinds are beefing up their bottom lines even as they shed physical assets. Internet companies are obvious examples, what with their high-flying stock prices and small or non-existent traditional assets.

Indeed, many stories in the media convey worry and awe at how many company stocks "levitate" on this alleged cloud called intellectual capital. Yet to some these supposedly skewed stock prices may be justified with a different valuation model. We believe that the valuation of a company based on intellectual capital is fundamentally the correct approach; however, robust metrics need to be developed that rationalize time-tested measurements of return on equity with more ephemeral estimates of the present value of future efficiencies due to technological advances.

Figuring out how much capital your company should have isn't easy. As a small organization, if your company's market cap is twice book value, your intellectual capital is roughly at parity with the Micro Cap 50 company average (on average 52 percent of the market cap of the Micro Cap 50 is attributable to intellectual capital). Not bad, but you could do better. In fact, for the NASDAQ 100 companies and the Dow Industrial components, the percent of the average market cap attributable to intellectual capital is 70 percent and 83 percent, respectively.

A look at the growth of market capitalization of the corporations that make up the Dow (Table 6-2), the NASDAQ 100 (Table 6-3) and even the Micro Cap 50 is testimony to the awakening of corporations to the importance of intellectual capital and its place in the determination of corporate value. However, parceling out increases in company valuation due to intellectual capital vs. market enthusiasm is difficult at best. While economies once were driven by wars or even weather, the global market runs on the power of ideas.

**Table 6-2. Intellectual capital of the Dow components** (October 1999)

| Company | Market Capitalization (MC) ($) | Book Value ($) | Intellectual Capital (IC) ($) | IC as % of MC |
|---|---|---|---|---|
| Coca-Cola Co. | 149,433,790,000 | 9,460,023,400 | 139,973,766,600 | 94 |
| Procter & Gamble Co. | 145,215,900,000 | 10,565,926,800 | 134,649,973,200 | 93 |
| Microsoft Corp. | 443,762,150,000 | 30,444,177,000 | 413,317,973,000 | 93 |
| Merck & Co. | | | | |
| Home Depot, Inc. | 121,979,550,000 | 10,025,282,800 | 111,954,267,200 | 92 |
| General Electric Co. | 451,293,070,000 | 39,987,330,000 | 411,305,740,000 | 91 |
| Wal-Mart Stores, Inc. | 258,582,490,000 | 22,955,446,800 | 235,627,043,200 | 91 |
| Intel. Corp. | 266,862,380,000 | 29,166,930,000 | 237,695,450,000 | 89 |
| Johnson & Johnson | 146,270,230,000 | 15,908,006,400 | 130,362,223,600 | 89 |
| International Bus. Mach. | 187,358,640,000 | 19,828,600,000 | 167,530,040,000 | 89 |
| American Express Co. | 69,388,850,000 | 9,745,775,900 | 59,643,074,100 | 86 |
| SBC Communications, Inc. | 175,686,070,000 | 26,506,422,600 | 149,179,647,400 | 85 |
| McDonald's Corp. | 64,146,650,000 | 9,491,680,200 | 54,654,969,800 | 85 |
| Minnesota Mining & Mfg. | 41,061,280,000 | 6,369,107,100 | 34,692,172,900 | 84 |
| AlliedSignal, Inc. | 33,536,190,000 | 5,421,006,000 | 28,115,184,000 | 84 |
| DuPont(E.I.) deNemours | 63,781,530,000 | 10,674,994,400 | 53,106,535,600 | 83 |
| Eastman Kodak Co. | 20,711,250,000 | 3,900,816,000 | 16,810,434,000 | 81 |
| Hewlett-Packard Co. | 95,392,640,000 | 19,012,300,800 | 76,380,339,200 | 80 |
| Citigroup Inc. | 190,499,360,000 | 44,708,344,200 | 145,791,015,800 | 77 |
| Exxon Corp. | 193,465,730,000 | 43,797,331,600 | 149,668,398,400 | 77 |
| Alcoa, Inc. | 23,381,530,000 | 5,899,201,000 | 17,482,329,000 | 75 |
| Philip Morris | 60,632,340,000 | 15,805,815,200 | 44,826,524,800 | 74 |
| United Technologies Corp. | 27,318,400,000 | 7,104,501,600 | 20,213,898,400 | 74 |
| Boeing Co. | 40,828,180,000 | 12,158,365,400 | 28,669,814,600 | 70 |
| Caterpillar, Inc. | 17,440,590,000 | 5,396,555,400 | 12,044,034,600 | 69 |
| General Motors | 44,654,650,000 | 16,530,222,200 | 28,124,427,800 | 63 |
| Walt Disney Co. | 55,167,270,000 | 20,726,416,500 | 34,440,853,500 | 62 |
| International Paper Co. | 1,318,290,000 | 536,141,200 | 782,148,800 | 59 |
| J.P. Morgan & Co. | 24,001,530,000 | 11,202,158,000 | 12,799,372,000 | 53 |
| AT&T Corp. | 148,785,080,000 | 74,771,190,000 | 74,013,890,000 | 50 |
| | | | Average= | 80 |

**Table 6-3. Intellectual capital of the NASDAQ 100 companies** (November 1999)

| Company | Market Capitalization (MC) ($) | Book Value ($) | Intellectual Capital (IC) ($) | IC as % of MC |
|---|---|---|---|---|
| 3Com Corporation | 10,005,840,000 | 3,236,381,400 | 6,769,458,600 | 68 |
| Adaptec, Inc. | 4,649,180,000 | 739,771,200 | 3,909,408,800 | 84 |
| ADC Telecommunications, Inc. | 6,483,520,000 | 1,070,005,200 | 5,413,514,800 | 83 |
| Adobe Systems Incoporated | 8,406,340,000 | 573,354,000 | 7,832,986,000 | 93 |
| Altera Corporation | 9,670,390,000 | 1,024,232,000 | 8,646,158,000 | 89 |
| Amazon.com, Inc. | 23,815,030,000 | 573,240,000 | 23,241,790,000 | 98 |
| American Power Conversion Corporation | 4,312,140,000 | 761,032,800 | 3,551,107,200 | 82 |
| Amgen Inc. | 40,668,990,000 | 2,784,381,600 | 37,884,608,400 | 93 |
| Andrew Corporation | 1,056,380,000 | 465,223,500 | 591,156,500 | 56 |
| Apollo Group, Inc. | 2,019,520,000 | 219,505,000 | 1,800,015,000 | 89 |
| Apple Computer, Inc. | 12,890,510,000 | 2,833,096,800 | 10,057,413,200 | 78 |
| Applied Materials, Inc. | 33,986,680,000 | 3,666,889,800 | 30,319,790,200 | 89 |
| At Home Corporation | 13,750,410,000 | 7,619,209,000 | 6,131,201,000 | 45 |
| Atmel Corporation | 3,871,540,000 | 770,304,000 | 3,101,236,000 | 80 |
| Autodesk, Inc. | 1,143,770,000 | 644,160,000 | 499,610,000 | 44 |
| Bed, Bath & Beyond Inc. | 4,658,460,000 | 469,862,400 | 4,188,597,600 | 90 |
| Biogen, Inc. | 11,142,170,000 | 867,346,400 | 10,274,823,600 | 92 |
| Biomet, Inc. | 3,398,760,000 | 798,765,600 | 2,599,994,400 | 76 |
| BMC Software, Inc. | 15,358,010,000 | 1,361,446,300 | 13,996,563,700 | 91 |
| Cambridge Technology Partners, Inc. | 968,483,750 | 259,240,000 | 709,243,750 | 73 |
| CBRL Group, Inc. | 710,900,000 | 790,918,700 | -80,018,700 | -11 |
| Chiron Corporation | 5,169,280,000 | 1,608,912,200 | 3,560,367,800 | 68 |
| CIENA Corporation | 4,842,080,000 | 516,473,600 | 4,325,606,400 | 89 |
| Cintas Corporation | 6,693,230,000 | 916,492,500 | 5,776,737,500 | 86 |
| Cisco Systems, Inc. | 290,600,000,000 | 11,674,920,000 | 278,925,080,000 | 96 |
| Citrix Systems, Inc. | 5,632,230,000 | 378,547,300 | 5,253,682,700 | 93 |
| CMCI, Inc. | 10,430,970,000 | 690,997,500 | 9,739,972,500 | 93 |
| CNET, Inc. | 3,440,010,000 | 250,776,000 | 3,189,234,000 | 93 |
| Comair Holdings, Inc. | 2,206,970,000 | 441,810,600 | 1,765,159,400 | 80 |
| Comcast Corporation | 29,845,690,000 | 6,156,865,000 | 23,688,825,000 | 79 |
| Compuware Corporation | 9,932,220,000 | 849,921,800 | 9,082,298,200 | 91 |
| Converse Technology, Inc. | 8,090,050,000 | 486,129,600 | 7,603,920,400 | 94 |
| Concord EFS, Inc. | 5,534,490,000 | 658,490,000 | 4,876,798,100 | 88 |
| Conexant Systems, Inc. | 9,101,080,000 | 952,281,900 | 8,148,798,100 | 90 |
| Costco Wholesale Corporation | 17,734,880,000 | 3,367,505,000 | 14,367,375,000 | 81 |

*continued*

**Table 6-3. Intellectual capital of the NASDAQ 100 companies** (November 1999) *(continued)*

| Company | Market Capitalization (MC) ($) | Book Value ($) | Intellectual Capital (IC) ($) | IC as % of MC |
|---|---|---|---|---|
| Dell Computer Corporation | 102,395,570,000 | 3,394,040,300 | 99,001,469,700 | 97 |
| Dollar Tree Stores, Inc. | 2,698,290,000 | 282,446,400 | 2,415,843,600 | 90 |
| eBay Inc. | 17,379,780,000 | 839,888,600 | 16,539,891,400 | 95 |
| Electronic Arts Inc. | 5,035,940,000 | 688,012,800 | 4,347,927,200 | 86 |
| Electronics for Imaging, Inc. | 2,212,260,000 | 472,516,800 | 1,739,743,200 | 79 |
| Fastenal Company | 1,375,290,000 | 265,959,400 | 110,933,060,000 | 81 |
| First Health Group Corp. | 1,159,970,000 | 107,762,400 | 1,052,207,600 | 91 |
| Fiserv, Inc. | 3,924,030,000 | 935,955,300 | 2,988,074,700 | 76 |
| Genzme General | 3,188,600,000 | 831,932,800 | 2,356,667,200 | 74 |
| Global Crossing Ltd. | 15,053,770,000 | 834,758,400 | 14,219,011,600 | 94 |
| Herman Miller, Inc. | 1,735,560,000 | 263,265,800 | 1,472,294,200 | 85 |
| Immunex Corporation | 10,308,310,000 | 305,969,400 | 10,002,340,600 | 97 |
| Intel Corporation | 256,164,900,000 | 25,471,600,000 | 230,693,300,000 | 90 |
| Intuit Inc. | 5,515,960,000 | 1,524,589,500 | 3,991,370,500 | 72 |
| JDS Uniphase Corporation | 28,911,590,000 | 3,896,392,500 | 25,015,197,500 | 87 |
| KLA-Tencor Corporation | 7,091,760,000 | 1,244,884,000 | 5,846,876,000 | 82 |
| Level 3 Communications, Inc. | 23,239,980,000 | 3,613,030,700 | 19,626,949,300 | 84 |
| Lincare Holdings Inc. | 1,637,800,000 | 545,615,100 | 1,092,184,900 | 67 |
| Lincare Technology Corporation | 10,785,980,000 | 909,898,000 | 9,876,082,000 | 92 |
| LM Ericsson Telephone Company | 76,407,080,000 | 7,613,898,000 | 68,793,182,000 | 90 |
| Lycos, Inc. | 4,657,450,000 | 376,090,600 | 4,281,359,400 | 92 |
| Maxim Integrated Products, Inc. | 10,771,170,000 | 882,831,500 | 9,888,338,500 | 92 |
| MCI WORLDCOM, Inc. | 160,737,190,000 | 48,401,162,400 | 112,336,027,600 | 70 |
| McLeodUSA Incorporated | 6,718,560,000 | 742,260,800 | 5,976,299,200 | 89 |
| Microchip Technology Incorporated | 3,387,750,000 | 389,002,500 | 2,998,747,500 | 89 |
| Micron Electronics, Inc. | 985,800,000 | 444,351,600 | 541,448,400 | 55 |
| Microsoft Corporation | 475,913,410,000 | 27,609,908,700 | 448,303,501,300 | 94 |
| Molex Incorporated | 5,723,530,000 | 1,499,103,600 | 4,224,426,400 | 74 |
| Network Associates, Inc. | 2,545,290,000 | 572,638,800 | 1,972,651,200 | 78 |
| Nextel Communications, Inc. | 27,043,300,000 | 1,553,161,500 | 25,490,138,500 | 94 |
| Northwest Airlines Corporation | 2,480,000,000 | -255,429,000 | 2,735,429,000 | 110 |
| Novell, Inc. | 6,682,300,000 | 1,548,775,500 | 5,133,524,500 | 77 |
| NTL Incorporated | 7,668,280,000 | 1,322,620,000 | 6,345,660,000 | 83 |

*continued*

**Table 6-3. Intellectual capital of the NASDAQ 100 companies** (November 1999) *(continued)*

| Company | Market Capitalization (MC) ($) | Book Value ($) | Intellectual Capital (IC) ($) | IC as % of MC |
|---|---|---|---|---|
| Oracle Corporation | 67,714,110,000 | 3,502,228,200 | 64,211,881,800 | 95 |
| PACCAR, Inc. | 3,689,790,000 | 1,961,415,000 | 1,728,375,000 | 47 |
| PacifiCare Health Systems, Inc. | 2,480,000,000 | 2,333,417,300 | 146,582,700 | 6 |
| PanAmSat Corporation | 5,895,850,000 | 2,750,861,800 | 3,144,988,200 | 53 |
| Parametic Technology Corporation | 5,132,850,000 | 444,279,000 | 4,688,571,000 | 91 |
| Paychex, Inc. | 9,709,520,000 | 463,589,200 | 9,245,930,800 | 95 |
| PeopleSoft, Inc. | 3,648,710,000 | 479,202,500 | 3,169,507,500 | 87 |
| QUALCOMM Incorporated | 35,761,400,000 | 1,409,629,000 | 34,351,771,000 | 96 |
| Quintiles Transitional Corp. | 2,130,270,000 | 896,275,600 | 1,233,994,400 | 58 |
| Qwest Communications International Inc. | 26,845,200,000 | 6,472,676,000 | 20,372,524,000 | 76 |
| Reuters Group PLC | 13,007,790,000 | 905,404,500 | 12,102,385,500 | 93 |
| Rexall Sundown, Inc. | 680,220,000 | 219,744,000 | 460,476,000 | 68 |
| Ross Stores, Inc. | 1,865,960,000 | 436,065,400 | 1,429,894,600 | 77 |
| Sanmina Corporation | 2,253,650,000 | 566,384,300 | 4,687,265,700 | 89 |
| Siebel Systems, Inc. | 10,169,890,000 | 425,079,900 | 9,744,810,100 | 96 |
| Sigma-Aldrich Corporation | 2,794,000,000 | 1,316,745,000 | 1,477,255,000 | 53 |
| Smurfit-Stone Container Corporation | 4,693,470,000 | 1,558,347,200 | 3,135,122,800 | 67 |
| Staples, Inc. | 10,305,130,000 | 1,741,687,500 | 8,563,442,500 | 83 |
| Starbucks Corporation | 4,953,360,000 | 933,357,600 | 4,020,002,400 | 81 |
| Stewart Enterprises, Inc. | 561,300,000 | 1,141,350,000 | -580,050,000 | -103 |
| Sun Microsystems, Inc. | 82,592,660,000 | 4,831,604,500 | 77,761,055,500 | 94 |
| Synopsys, Inc. | 4,414,560,000 | 853,742,500 | 3,550,817,500 | 80 |
| Tech Data Corporation | 1,385,000,000 | 970,173,000 | 414,827,000 | 30 |
| Tellabs, Inc. | 24,811,390,000 | 1,616,193,600 | 23,195,196,400 | 93 |
| USA Networks, Inc. | 7,526,600,000 | 2,663,969,000 | 4,862,631,000 | 65 |
| VERITAS Software Corporation | 18,307,140,000 | 3,582,578,100 | 14,724,561,900 | 80 |
| VISX, Incorporated | 3,988,770,000 | 221,247,200 | 3,767,522,800 | 94 |
| Vitesse Semiconductor Corporation | 7,040,440,000 | 448,132,400 | 6,592,307,600 | 94 |
| Voice Stream Wireless Corporation | 9,440,400,000 | 205,540,000 | 9,234,860,000 | 98 |
| Worthington Industries, Inc. | 1,481,500,000 | 686,147,000 | 795,353,000 | 54 |
| Xilinx, Inc. | 12,427,310,000 | 937,295,800 | 11,490,014,200 | 92 |
| Yahoo! Inc. | 46,376,420,000 | 813,260,000 | 45,563,160,000 | 98 |
| | | | Average= | 79 |

# public venture capital— thirst for the new idea

# 7

*Great ideas need landing gear as well as wings*
—C.D. JACKSON

YOU'VE HEARD THE OLD SAYING THAT IT TAKES MONEY TO MAKE money. This saying is especially true when it comes to launching a company. When the company is a technology startup, the saying is truer still.

Entrepreneurs are hungry for fresh technologies to develop and sell. If they already have the technology—stuff that has "blockbuster product" written all over it—then the need is for management and capital. Without cash or some other sort of currency, entrepreneurs cannot acquire technologies or spin great ideas into greater products.

Conversely, Wall Street is perpetually on the lookout for fresh investment ideas, combined with management teams that can implement them. Investment bankers are continuously turning over rocks in the hunt for the next big thing, which is currently the hot new Internet technology company of the moment. As the human genome becomes better understood, companies that focus on it and other "nano" technologies are not far behind. Indeed, as the

FIGURE 7-1
Money-tree survey—technology companies
as a percentage of total investments for (1995–YTD Q2 1999)

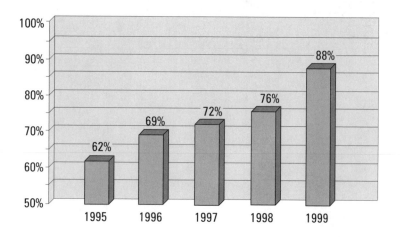

It is important to distinguish that these figures are only cash-for-equity transactions in private companies. They do not include loans, subordinated debt, stock-for-equity, secondary purchases or other financial instruments. As such, these figures represent only a small portion of the overall private equity market in the U.S.

Source: Price Waterhouse Coopers MoneyTree™ Venture Capital Survey

world shifts to a knowledge economy, investors increasingly move cash into technology ventures.

And it shows. Technology corporations received 88 percent of investment capital during offerings on public stock exchanges in the first six months of 1999. In the second quarter of 1999, the amount of U.S. capital chasing technology companies was about 87.7 percent of the money invested in private companies during this period (approximately $10.5 billion). The reason is simple. Capital markets are sensitized to the potentially massive profits that can emanate from companies armed with powerful intellectual capital and sound cutting-edge technologies—especially when these companies have proven management talent.

Of course, buying good technology and/or turning it into products and services takes money, often lots of it. Small publicly owned companies, even ones with smart management and keen employees, often lack the resources to find, acquire, refine and market the technologies they need to strengthen their core competencies.

Friends and family members might be a good source of investment dollars, and plenty of successful ventures have gotten started on the strength of those small personal loans. But this path often does not generate enough money to get the company started properly. Plus, there is always the danger of souring important personal relationships should the business fail.

Another source is wealthy private investors, or "angels." Or there is the possibility of teaming up with another company in a joint venture. This is a direction that has become increasingly popular among medical and biotechnology companies; joint ventures that benefit from having partners with financial depth and marketing expertise.

Then there is the investment banker or venture capitalist who uses money pooled from various sources to invest in private or public corporations. These people provide seed money and funds for expansion. Many venture capitalists specialize in technology sectors such as the Internet, pharmaceuticals and telecommunications.

Unlike other types of investors, venture capitalists are rarely passive financiers, placing their money in a company and waiting for equity appreciation. Rather, they typically work to varying degrees as partners who invest money (from private individuals, corporate pension funds, etc.) and become involved as advisors to management.

Venture capitalists meet with a company's management to size up executives as people and managers. Indeed, the quality of a company's management may be the most important key to winning venture funding. If you were investing millions of dollars in a company, you'd want to know if the folks running it are capable of building a

$70 million company. You'd want to know if they'd ever successfully started and built a company before. You'd like to feel certain that the top executives are also able to attract other talent.

What venture capitalists also look for in a candidate for investment is a company that has a significant advantage over current or potential competitors. They don't want companies that make a slightly better widget; they want an outfit that has a product that will allow it to potentially dominate a market segment or industry.

Companies with patented proprietary technologies may be more attractive to venture capitalists because these companies have taken steps to exclude competitors; i.e., have established barriers to entry. Companies can also rely on trade secret protections, but these companies typically need a nice head start on would-be competitors to entice venture capitalists (Lister and Harnish, 1996).

Venture capitalists also look for companies with significant growth potential. If you blink at the notion that your company is expected to grow from peanuts to $30 million, $50 million or even $100 million in five to seven years, then you'll likely be passed over quickly.

The size of the investment that venture capitalists make varies significantly. Some funds won't invest less than $20 or $50 million, often in companies with "proven" track records. Others will ante up $4 to $9 million or so on a smattering of different small companies. Most avoid taking stakes smaller than $1 million. Time, after all, is more precious than money.

But one thing that most venture capitalists share is a craving for big returns. Indeed, most venture capitalists turn up their noses at deals that promise annual returns smaller than 20 percent or so. They won't consider a deal that doesn't earn returns at "sell-out" of between 20 and 50 percent per year.

Business trends also play a big role in what types of companies venture capitalists and investment bankers are interested in. An especially hot type of Internet technology, for example, may spawn many similar businesses that are able to acquire funding, too, in the rush to invest in that moment's new thing. You could barely get a venture capitalist to glance at entertainment media companies a few years ago; these days, the press is packed with stories about Internet, interactive television and similar IPOs (Lister and Harnish, 1996).

Of course, not all bankers like the same types of companies. For example, some prefer certain industries, technologies or even geographical areas. Some stay close to home. (It's a lot easier to invest in companies in your backyard than it is to take time to scour the nation and/or globe.) Others fancy searching for plum investments in far-flung locales (a detailed list of investment banks is contained in Appendix III).

And, because venture capitalists and investment bankers are human, they often pick industries and types of companies to invest in based in part on personal preferences. For example, if your business plan for a novel medical instrument lands on the desk of a partner at a venture capital firm who happens to be big into healthcare industry deals, you may stand a better chance that your plan will be passed along to the right partner, not pitched in the waste basket.

Some even favor investing in companies at a specific age in a business' life cycle. Some want to be in on the ground floor of a new venture's life; others prefer to invest in juvenile companies. And venture capitalists tend to invest in rounds of financing—chunks of cash, rather than one massive lump sum—anteing up fresh funding when certain benchmarks of performance are met.

As you can tell, enticing investors to help your company buy a core technology, then add additional cash to turn that technology into products and (hopefully) profits, is no easy task.

Consider also that there are more than 5,500 public companies out there, each with market capitalization of less than $250 million.

That's a constellation of potential star companies that may never fulfill their corporate destinies for lack of money to acquire and develop technologies. In many cases the core competencies of these companies lie outside of the technology development realm.

In the next chapter, we will show how these companies can help to make their weaknesses irrelevant by teaming up with university research teams, coupled with equity financing.

# the new mortgage— financing the arbitrage between university technolology and small public company securities

## 8

*What we need are more people who specialize in the impossible*

—T HEODORE  R OETHKE

Y OU'VE SEEN HOW COMPANIES AND UNIVERSITIES HAVE FINANCED technology transfer in the past. It has been a system of count- less potentially brilliant opportunities missed, punctuated by rare sparks of synergy and success. Now, let's examine a novel, improved way for small- and micro-cap companies to identify, acquire and develop the best technologies from the nation's top idea factories.

The first step is to assemble a database cataloging the thou- sands of patented technologies developed at universities and research labs around the nation and another cataloging the many micro- and small-cap public companies. Some of this information can be mined from various Internet sites, including sites of the U.S. Patent Office and the Securities Exchange Commission. Of course, forging personal relationships with university technology transfer directors, as well as scientists and researchers, is paramount. These human connections can mean the difference between early alerts on cutting-edge technologies and missed opportunities. Indeed, to be successful, it is necessary to build a network of carefully nurtured personal relationships. Instead of having inventory, you must have access and knowledge. You need to know what technologies are being developed, roughly how much it will cost to continue their development and what additional monies are need to acquire the licenses.

Work on these lists of technologies and companies must be ongoing, since new ventures and inventions occur daily. You will never be able to say, "Well, that's done," nor will you want to.

Learning begins with asking lots of questions. What are you good at? What are your weaknesses? What technologies do you currently have and how are you using them?

Listen way beyond your capacity for patience. Listen—it's the least expensive way to figure out what makes your business tick. Otherwise, you won't really know what you need, and you might miss managerial strengths or flaws that cannot be seen from the bridge or in the financial statements. If there is one thing you don't want to do, it's to transfer technology to a company that lacks the ability and focus to bring it to market successfully. In an economy that increasingly rewards nimble, smart moves, a near miss for a small company can be fiscally fatal.

After you understand your company's technology strengths and weaknesses, you need to set to work shopping for technologies that you think offer a good match for your current core technology or talents. A word of warning: You may find technology at a university that at first review seems to be a perfect fit for your company, but the match may not ultimately work out. For example, a particular technology that seems complementary to your company's mission and methods may use a process that is at odds with the way your company does things. If you've listened well, and if your managers have presented a clear picture, you should be able to discern which technologies will fit and which won't. There is an art to saying "No." Practice it.

It's important to meet the folks who invented the science you want to use. A broad network of contacts can help open doors, or you may already enjoy a close relationship with the university research team and/or the prospective technology's inventor.

Acquiring a technology is not like buying a book. You can't decide whether to plunk down $25 based on reading a paragraph or two. You want to get to know the scientists and researchers to learn how their technology works. You want to find out what the technology can accomplish and what types of products and services it might yield. You can't very well present technology to your team and to your customers if you haven't a clue how the stuff works.

To this end, it is beneficial to assemble a "kitchen cabinet" (i.e., a scientific advisory council comprised of relevant scientists and researchers) to help you discern promising technologies from passed over technologies.

What happens once you find technology that fits your needs? Seek to license it, not purchase it. Licensing is a far better way to maintain the university alliance for technology outsourcing, research and development and the development of improvements. Keeping the university in the loop will greatly increase the half-life of the technology and improve its long-term value by leveraging the university's structural intellectual capital.

For a small public company, one of the best ways to finance the acquisition of a new technology is to use unregistered common shares for the up-front license fee. This swap of stock for technology does several things. It allows you to conserve cash—a precious commodity—and it makes the university a potential long-term technology investor.

This type of deal—an equity swap, in essence—solves what has been the biggest stumbling block for small companies trying to acquire university technology. Of course, small companies had the currency all along, but until recently few universities, or companies, were willing to exchange it for technology licenses.

Think about it. The one thing small public companies are long in is stock. They rarely have much free cash. And being publicly owned isn't cheap—there are legal fees, accounting costs, investor relations fees, etc.

But, by using a company's stock as a technology license currency, the company has found a currency to trade for what it really needs, without saddling itself with painful and potentially debilitating debt—and it has simultaneously boosted the company's intellectual capital, making its own common stock potentially more valuable.

The benefits of this method of arbitrage should be clear to most companies. Curiously, few CEOs and university technology transfer directors have used it before. Sometimes a measure of the value of an idea is how obvious it seems in retrospect.

Universities and inventors benefit greatly with this model as well. Universities and inventors will get up-front license fees much more frequently than they otherwise would. That is in stark contrast to most technology transfer arrangements, in which universities share royalties with technology transfer companies, and inventors typically take home fractional earnings. What's more, should the client company fail to pay required royalties to the university; the technology licenses revert to the university.

It's important to note that this new method of technology transfer is built for speed. License agreements, coupled with sponsored research, can be completed in a matter of weeks to months instead of the traditional one or two years during which companies may wind up with stale technology and frustrated research teams. The benefits of being small are clarity of purpose and swiftness in action. Use your strengths.

Acquiring technology is only the first step for a company. A new technology that could solve big problems and create a significant earning stream in the process is great, but even potentially decisive technology is grounded without further refinement. That is why it is imperative that companies work with universities throughout the process of turning technology into products.

This often means helping foster better links to the places and people who invented the technology. This isn't handholding. It's good business. If you have invested significant equity to acquire a technology license and help fund continuing research for a company, and your university resource has invested in the technology development and has a vested interest in your equity, together you both have a very real interest in mutual success. Indeed, this entire approach fosters greater cooperation between universities and small companies. Sharing knowledge (in the form of access to each other's network of contacts, expertise, etc.) makes both smarter, and each is enriched by fresh interchange of intellectual capital.

Suggesting that companies tap the knowledge of the folks who invented their technology may seem obvious, but companies constantly need new knowledge that they can germinate into fresh products, and old habits of technology transfer die hard. Also, universities and inventors are often unaccustomed to working with companies, let alone small ones. Yet it is vital for companies and universities everywhere to form bridges between one another. This book isn't about tackling one project; it is about fostering greater linkage between these idea factories and the marketplace. So far, universities have embraced this new method of technology transfer, which in effect offers them an off-balance-sheet technology marketing program without having to share royalties.

After a company has used this new model for acquiring and developing university technology, the traditional approach of doing research in-house becomes pretty unattractive. In any case, companies need time to nurture their technologies. You could say that this model helps create a public company incubator for embryonic university technology products and services.

Sometimes help is needed. In an effort to create an efficient market, facilitators are often necessary to reduce transaction time and information exchange. Technology brokerage, when done properly, has a vested interest in both the university's and company's futures. When a company acquires a powerful new technology that matches its mission and its core technology, it has a chance to become a stronger company. From the university perspective, this is an opportunity to move a model, conjecture or hypothesis into the market. It gives legs to ideas.

As we move into a knowledge economy, increasing the amount of intellectual capital you can access and leverage means you have boosted your company's chances of survival—and success. Universities benefit, too, from this arbitrage in much the same way as the companies they are connected to. They need a bridge to the marketplace—even as they learn new ways of doing things.

# creating linkage—a bridge is a series of links

**9**

*You don't get harmony when*
*everyone sings the same note*

— DOUG FLOYD

WE HAVE INTRODUCED SOME OF THE IMPORTANT BENEFITS OF linking universities and industry. But it is just as important to note that these two cultures should remain essentially separate. We want to create a bridge to connect them; we don't want to merge them. Universities and industry can learn from each other without converting one another.

Scientists and researchers typically function best in academia or in research environments, while business types thrive in the commercial world. Universities aren't commercial entities, nor should they be. To make them into things that they are not would distort their missions and compromise their creativity.

Just imagine how many CEOs would have indulged Albert Einstein's scientific whims. ("Say, Albert, what's all this I hear about you spending our time and money on something called relativity?") Yet this man changed our understanding of physics and, in the process, enlarged our world.

What business would have had the foresight, not to mention the formidable wallet, to fund Einstein's scientific musings? And it is unlikely that a "pure" scientist like Einstein would have felt comfortable in an applied research center. In fact, in his early years, he eschewed even working in universities because he viewed them as too dogmatic in their need to rapidly publish results (i.e., form over substance). Not to mention that they wouldn't hire him.

This is not to say that scientists shouldn't become business folks, or that executives have no business in academia. Plenty of both types have become cross-cultural successes. But in much the same way that companies do best when they are able to leverage the talents and knowledge of their networks rather than own the means of production, so it is that universities and companies work together best in a symbiotic but separate fashion.

This is why technology transfer is really about building bridges, and not about extracting scientists and putting them into commercial enterprises. It is, rather, about extracting their ideas.

Scientists may be motivated more by tackling big questions than by solely monetary rewards. Not that they or anyone else would disdain a decent income. In fact, they should be richly rewarded for their efforts through royalty sharing and other means of compensation, since they produce the ideas that generate wealth. Their seeds are used to germinate companies and nourish the economy.

To be sure, the unique (if so far limited) interplay between universities and industry in America has helped catapult the nation to the top of the global economy. Many of the biggest advances in medicine, computing and other sciences have been fueled in large part by federal funds. Cold War funding spurred us to make great strides. Money used for space exploration has yielded countless innovations. It was the Department of Defense that funded the creation of the once-quirky idea of linking computers together into what is now known as the Internet, a technology that is revolutionizing the way the world lives and works. Indeed, federal funding of science is an investment in our future and the future of the planet.

Federal and state funding also reaps less obvious rewards. Inventors and business people who leverage university science into commercial products and services may plow money back into the university itself in the form of further funding of research and development. Or they may donate money to the school in appreciation of the university's contribution to their development and success.

Our model for growing and transferring technology doesn't just help enrich a handful of investors, business folk and scientists. It is a tool for tapping a source of tremendous knowledge that the entire world can use to solve big problems of health and survival, and to improve the standard of living for all people around the world.

Is there a more appropriate way for federal and state taxes to be used than to help strengthen local and global economies? Nowhere else do people enjoy such a high standard of living, wealth and opportunity as in United States. Yet we also have problems. With great shifts in the economy, many are cast aside. Poverty often begets more of the same. Further, more than two hundred thousand Americans die every year from infectious diseases and cancer. AIDS is but one of a constellation of potential viral catastrophes that threaten our existence. Who knows what opportunistic organisms (i.e., plague) we'll have to face tomorrow—and whose outcome will be dependent upon a few dedicated teams such as those that precisely and methodically unraveled the human genome decades before.

As the world's population continues, at least in the near term, to boom, so too do problems such as deforestation and pollution. We face formidable challenges in the coming years to protect and preserve not just the trees and land, but the amazing bio-diversity of our planet. Not counting the ravages of pollution and depletion of natural habitats, it is estimated that more than a quarter of known animal extinctions have been caused by human hunting.

The race to create antidotes to future diseases and to fix our myriad other problems (most of which we have yet to face) will be won by using our collective brainpower—knowledge that can be used to create powerful medicines and tools. Innovation can lead to new products and services that save lives, reduce suffering and improve the quality of life.

In working to improve the standard of living in the United States and the world, it is necessary to encourage further technological development. Part of moving forward means using what we have, and currently we make poor use of the "knowledge factory" capacity of the universities. Just imagine how much faster our growth and development would be if we made full use of the ideas and knowledge that flow out of our universities and government laboratories every single day.

The faster we link the research capabilities of our frequently ignored universities to the smaller public companies that need it, the more we invigorate these companies and the economy with new ideas. That's why these idea factories are an essential engine for economic development and growth in the 21st century.

Of course, our rush to innovate, to pierce the secrets of nature, must be tempered with a deep sense of responsibility—to fellow humans and to the living world we share. There must be a balance between our strength and our sensibility or we, to paraphrase Jose Delgado at Yale University, can inadvertently select ourselves for extinction.

Innovation for innovation's sake often has led to unintended, if ancillary, consequences. Splitting the atom unleashed forces that were later used in two atomic bombs exploded during World War II. Rocket technology was used to create the V-2 rockets that rained down on Londoners during the Blitzkrieg. Advances in chemistry were turned into the poison gas that was used on French troops during World War I. The litany of awful applications and misuse of science goes on and on.

Yet technology is neither good nor evil. How we use it makes it so. This is why the responsible use of science—and technology transfer—is paramount.

Strengthening links between universities and industry makes more than fiscal sense. It promises to better engage some of our brightest and most dedicated people. It also satisfies our urge to create—the spark that makes humans more than mere carbon-based, ungrateful bipeds.

# bibliography

Association of University Technology Managers, Inc. AUTM *Licensing Survey, Fiscal Year* 1998.

Battelle. *Technology Forecasts* (www.battelle.org), 1999.

Brooking, A. *Intellectual Capital*. International Business Press, London, 1996.

Brown, Lester, Michael Renner and Brian Halweil. *Vital Signs*, Linda Starke (ed.) New York, NY: W.W. Norton & Company, 1999.

Drucker, P.F. *Management Challenges for the 21st Century*. New York, NY: Harper-Collins, 1999.

Gross, C.M. *The Right Fit*. Portland, OR: Productivity Press, 1996.

Klein, D.A. *The Strategic Management of Intellectual Capital*. Woburn, MA: Butterworth-Heinemann, 1998.

Lister, Catherine E. and Thomas D. Harnish. *Directory of Venture Capital*. New York, NY: John Wiley & Sons, 1996.

Malone, M.S. and L. Edvinsson. "Intellectual Capital." *Realizing Your Company's True Value By Finding Its Hidden Roots*. New York, NY: HarperCollins, 1997.

Martin, M.J.C. *Managing Innovation and Entrepreneurship in Technology-Based Firms*. New York, NY: John Wiley & Sons, 1994.

Martins, P. "How Will Climate Change Affect Human Health?" *American Scientist*, November-December, 1999.

Moore, G. *Crossing the Chasm*. New York, NY: HarperCollins, 1991.

Mullis et al. *Mathematics Achievement in the Primary School Years*, Boston College, 1997.

Muir, A. *The Technology Transfer System*. Latham, NY: Latham Book Publishing, 1997.

National Science Foundation. *Survey of Federal Funds for Research and Development: Fiscal Years 1997, 1998 , and 1999*, NSF 99-333. Arlington, VA: Division of Science Resources Studies, 1999.

Partington, A. (editor). *The Concise Oxford Dictionary of Quotations*. Walton Street, New York: Oxford University Press, 1993.

Stewart, T. "Intellectual Capital." *The New Wealth of Organizations*. New York, NY: Doubleday, 1997.

Sullivan, P. "Profiting from Intellectual Capital." *Extracting Value from Innovation*. Canada: John Wiley & Sons, 1998.

Svelby, K. "The New Organizational Wealth." *Managing and Measuring Knowledge-Base Assets*. San Francisco, CA: Berrett-Koehler, 1997.

The Economist. *Taking the Plunge*. October 18, 1997.

Ungson G. and J. Trudel. "Engines of Prosperity." *Templates for the Information Age*. London: Imperial College Press, 1998.

United States General Accounting Office. *University Research Controlling Inappropriate Access to Federally Funded Research Results*, Washington, D.C. 1992.

Walters, S.G. and D. Gray. "Managing the Industry/University Cooperative Research Center." *A Guide for Directors and Other Stakeholders*. Columbus, OH: Battelle Memorial Institute, 1998.

Washington Researchers. *Technology Opportunities*. Rockville, MD: Washington Researchers, 1999.

# U.S. university research expenditures and resulting technology production

APPENDIX

**Note:** Data on research expenditures and licensing income for the universities listed were derived from AUTM's Licensing Survey FY 1998. Data are used with permission. For additional information, please visit the Association of University Technology Managers, Inc. at www.autm.net. Also, please note that total research expenditures, total licensing income, invention disclosures received and licenses and options executed, while reported in a given year, are not contemporaneous. For example, an invention may be disclosed in 1998, but royalties received by a university in 1998 are not likely to be derived from disclosures made during the same year.

## Research Expenditures by U.S. Universities

| Name of Institution | FY 1998 Total Sponsored Research Expenditures | FY 1998 Invention Disclosures Received | FY 1998 Licenses & Options Executed | Percent of Invention Disclosures Licensed (%) | FY 1998 Gross License Income Received | FY 1998 Income as a Percentage of Research Expenditures |
|---|---|---|---|---|---|---|
| 1. Univ. of California System | $1,709,929,000 | 742 | 177 | 24% | $73,101,000 | 4.28% |
| 2. Johns Hopkins University | $987,463,936 | 228 | 93 | 41% | $5,513,284 | 0.56% |
| 3. Massachusetts Inst. of Technology (MIT) | $761,400,000 | 356 | 95 | 27% | $18,046,991 | 2.37% |
| 4. University of Michigan | $491,500,000 | 160 | 44 | 28% | $6,805,800 | 1.38% |

| Name of Institution | FY 1997 Total Sponsored Research Expenditures | FY 1997 Invention Disclosures Received | FY 1997 Licenses & Options Executed | Percent of Invention Disclosures Licensed (%) | FY 1997 Gross License Income Received | Income as a Percentage of Research Expenditures |
|---|---|---|---|---|---|---|
| 5. University of Minnesota | $432,928,695 | 144 | 65 | 45% | $3,199,373 | 0.74% |
| 6. Univ. of Washington/ Wash. Res. Fndtn. | $432,383,000 | 255 | 112 | 44% | $21,299,214 | 4.93% |
| 7. University of Pennsylvania | $414,356,000 | 233 | 72 | 31% | $7,246,695 | 1.75% |
| 8. Stanford University | $401,049,000 | 247 | 118 | 48% | $43,197,379 | 10.77% |
| 9. Texas A&M University System | $393,720,000 | 135 | 54 | 40% | $4,414,429 | 1.12% |
| 10. North Carolina State | $379,856,085 | 101 | 39 | 39% | $4,281,109 | 1.13% |
| 11. SUNY Research Foundation | $378,791,618 | 165 | 38 | 23% | $12,123,088 | 3.20% |
| 12. Harvard University | $374,446,700 | 124 | 44 | 35% | $8,877,826 | 2.37% |
| 13. Penn State University | $374,145,000 | 190 | 42 | 22% | $2,012,584 | 0.54% |
| 14. W.A.R.F./Univ. of Wisconsin-Madison | $362,100,000 | 208 | 75 | 36% | $16,121,075 | 4.45% |
| 15. Cornell Research Foundation, Inc. | $343,007,000 | 207 | 84 | 41% | $4,798,469 | 1.40% |
| 16. Univ. of Illinois, Urbana, Champaign | $338,841,000 | 104 | 34 | 33% | $3,060,496 | 0.90% |
| 17. University of Arizona | $302,327,804 | 90 | 32 | 36% | $477,000 | 0.16% |
| 18. Yale University | $299,800,000 | 65 | 45 | 69% | $33,261,248 | 11.09% |
| 19. University of Pittsburgh | $290,000,000 | 74 | 20 | 27% | $2,560,000 | 0.88% |
| 20. University of North Carolina/Chapel Hill | $283,699,269 | 106 | 66 | 62% | $1,868,030 | 0.66% |
| 21. Duke University | $282,000,000 | 112 | 49 | 44% | $1,318,680 | 0.47% |
| 22. Washington University | $265,315,567 | 34 | 36 | 106% | $4,548,313 | 1.71% |
| 23. University of Southern California | $262,473,088 | 122 | 73 | 60% | $983,642 | 0.37% |
| 24. Columbia University | $260,700,000 | 151 | 112 | 74% | $61,649,002 | 23.65% |
| 25. University of Texas at Austin | $244,843,000 | 91 | 23 | 25% | $1,455,599 | 0.59% |
| 26. University of Florida | $240,900,000 | 139 | 17 | 12% | $19,144,753 | 7.95% |
| 27. Georgia Institute of Technology | $220,096,991 | 115 | 18 | 16% | $2,305,729 | 1.05% |
| 28. University of Georgia | $217,945,000 | 88 | 94 | 107% | $3,383,662 | 1.55% |

| Name of Institution | FY 1997 Total Sponsored Research Expenditures | FY 1997 Invention Disclosures Received | FY 1997 Licenses & Options Executed | Percent of Invention Disclosures Licensed (%) | FY 1997 Gross License Income Received | Income as a Percentage of Research Expenditures |
|---|---|---|---|---|---|---|
| 29. University of Iowa Research Foundation | $217,321,805 | 90 | 22 | 24% | $900,596 | 0.41% |
| 30. Purdue Research Foundation | $216,479,000 | 73 | 28 | 38% | $870,000 | 0.40% |
| 31. Northwestern University | $214,022,093 | 94 | 15 | 16% | $1,324,555 | 0.62% |
| 32. Ohio State University | $209,685,896 | 75 | 16 | 21% | $1,758,533 | 0.84% |
| 33. Baylor College of Medicine | $207,100,000 | 86 | 32 | 37% | $7,247,178 | 3.50% |
| 34. Univ. of Massachusetts, All Campuses | $204,881,505 | 94 | 16 | 17% | $2,014,000 | 0.98% |
| 35. Michigan State University | $193,611,000 | 68 | 26 | 38% | $24,336,872 | 12.57% |
| 36. Indiana University (ARTI) | $193,305,225 | 61 | 25 | 41% | $1,007,882 | 0.52% |
| 37. University of Missouri System | $192,053,239 | 60 | 17 | 28% | $1,479,782 | 0.77% |
| 38. University of Utah | $183,908,056 | 197 | 32 | 16% | $2,170,498 | 1.18% |
| 39. University of Alabama/ Birmingham | $183,146,952 | 118 | 41 | 35% | $1,429,339 | 0.78% |
| 40. University of Rochester | $181,858,000 | 70 | 14 | 20% | $4,054,820 | 2.23% |
| 41. Iowa State University | $181,400,000 | 158 | 191 | 121% | $2,725,985 | 1.50% |
| 42. Carnegie Mellon University | $169,899,765 | 82 | 11 | 13% | $30,065,000 | 17.70% |
| 43. Case Western Reserve University | $169,068,664 | 52 | 7 | 13% | $1,244,000 | 0.74% |
| 44. Univ. of Texas Southwestern Med. Ctr. | $167,574,011 | 77 | 18 | 23% | $3,865,940 | 2.31% |
| 45. Virginia Tech Intellectual Properties, Inc. | $167,118,000 | 69 | 29 | 42% | $1,120,411 | 0.67% |
| 46. Emory University | $164,900,000 | 57 | 15 | 26% | $5,074,100 | 3.08% |
| 47. Univ. of Maryland, College Park | $164,290,433 | 102 | 69 | 68% | $1,797,713 | 1.09% |
| 48. Univ. of Virginia Patents Fndtn. | $163,701,079 | 117 | 6 | 5% | $3,751,868 | 2.29% |
| 49. University of Miami | $161,200,000 | 32 | 11 | 34% | $124,667 | 0.08% |
| 50. Univ. of Tennessee Research Corp. | $160,587,000 | 77 | 14 | 18% | $1,190,127 | 0.74% |

| Name of Institution | FY 1997 Total Sponsored Research Expenditures | FY 1997 Invention Disclosures Received | FY 1997 Licenses & Options Executed | Percent of Invention Disclosures Licensed (%) | FY 1997 Gross License Income Received | Income as a Percentage of Research Expenditures |
|---|---|---|---|---|---|---|
| 51. Univ. of Chicago– ARCH Dev. Corp. | $151,635,000 | 167 | 16 | 10% | $2,242,331 | 1.48% |
| 52. California Institute of Technology | $151,000,000 | 315 | 44 | 14% | $5,500,000 | 3.64% |
| 53. Vanderbilt University | $148,700,000 | 70 | 35 | 50% | $889,460 | 0.60% |
| 54. University of Hawaii | $147,043,269 | 20 | 4 | 20% | $942,774 | 0.64% |
| 55. Rutgers, The State University of NJ | $146,855,495 | 134 | 43 | 32% | $4,749,165 | 3.23% |
| 56. Colorado State University | $140,097,391 | 39 | 8 | 21% | $356,735 | 0.25% |
| 57. New York University | $139,013,000 | 46 | NA | NA | $2,500,000 | 1.80% |
| 58. Wayne State University | $138,425,000 | 37 | 8 | 22% | $1,053,000 | 0.76% |
| 59. Oregon State University | $138,240,000 | 28 | 16 | 57% | $632,974 | 0.46% |
| 60. University of Illinois at Chicago | $138,000,000 | 59 | 28 | 47% | $1,818,881 | 1.32% |
| 61. University of South Florida | $134,907,636 | 57 | 8 | 14% | $706,248 | 0.52% |
| 62. Boston University | $131,956,155 | 63 | 17 | 27% | $906,355 | 0.69% |
| 63. University of Maryland, Baltimore | $131,112,000 | 58 | 7 | 12% | $171,700 | 0.13% |
| 64. Albert Einstein College of Med./Yeshiva | $125,000,000 | 27 | 35 | 130% | $1,325,000 | 1.06% |
| 65. University of Nebraska– Lincoln | $118,857,000 | 26 | 6 | 23% | $766,179 | 0.64% |
| 66. Mount Sinai School of Medicine of NYU | $116,175,242 | 17 | 2 | 12% | $696,691 | 0.60% |
| 67. Florida State University | $112,077,647 | 14 | 7 | 50% | $46,642,688 | 41.62% |
| 68. Utah State University | $106,931,407 | 100 | 12 | 12% | $209,007 | 0.20% |
| 69. University of Kansas | $104,271,000 | 70 | 5 | 7% | $571,476 | 0.55% |
| 70. Univ. of Texas Houston Hlth. Sci. Ctr. | $103,753,489 | 14 | 5 | 36% | $341,959 | 0.33% |
| 71. University of Connecticut | $100,900,000 | 45 | 12 | 27% | $806,155 | 0.80% |
| 72. Oregon Health Sciences University | $94,624,306 | 56 | 13 | 23% | $449,687 | 0.48% |
| 73. Univ. of New Mexico/Sci. & Tech. Corp. | $94,600,000 | 45 | 3 | 7% | $366,000 | 0.39% |

| Name of Institution | FY 1997 Total Sponsored Research Expenditures | FY 1997 Invention Disclosures Received | FY 1997 Licenses & Options Executed | Percent of Invention Disclosures Licensed (%) | FY 1997 Gross License Income Received | Income as a Percentage of Research Expenditures |
|---|---|---|---|---|---|---|
| 74. Princeton University | $93,000,000 | 105 | 21 | 20% | $1,268,000 | 1.36% |
| 75. Clemson | $90,150,000 | 17 | 4 | 24% | $4,329,593 | 4.80% |
| 76. Auburn University | $88,472,824 | 31 | 7 | 23% | $192,319 | 0.22% |
| 77. Tulane University | $87,857,880 | 21 | 9 | 43% | $6,588,181 | 7.50% |
| 78. Washington State University | $84,923,259 | 61 | 22 | 36% | $241,000 | 0.28% |
| 79. Thomas Jefferson University | $84,000,000 | 76 | 15 | 20% | $797,858 | 0.95% |
| 80. Oklahoma State University | $82,200,000 | 15 | 2 | 13% | $137,634 | 0.17% |
| 81. Univ. of Kentucky Research Fndtn. | $80,706,598 | 33 | 11 | 33% | $2,361,958 | 2.93% |
| 82. Univ. of Texas Hlth. Sci. Ctr. San Antonio | $78,146,072 | 21 | 9 | 43% | $2,158,139 | 2.76% |
| 83. Dartmouth College | $76,313,745 | 19 | 9 | 47% | $618,777 | 0.81% |
| 84. University of Cincinnati | $75,511,132 | 38 | 7 | 18% | $3,269,292 | 4.33% |
| 85. University of Arkansas, Fayetteville | $75,461,203 | 18 | 1 | 6% | $206,546 | 0.27% |
| 86. Brown University Research Foundation | $73,977,000 | 26 | 5 | 19% | $350,898 | 0.47% |
| 87. University of Rhode Island | $72,000,000 | 15 | 10 | 67% | $873,664 | 1.21% |
| 88. Medical University of South Carolina | $68,745,282 | 55 | 5 | 9% | $46,870 | 0.07% |
| 89. New Mexico State University | $62,718,877 | 5 | 1 | 20% | $13,720 | 0.02% |
| 90. University of Delaware | $60,667,278 | 20 | 2 | 10% | $204,264 | 0.34% |
| 91. Wake Forest University | $57,202,499 | 49 | 9 | 18% | $1,741,880 | 3.05% |
| 92. Allegheny Univ. of the Health Sciences | $54,500,000 | 48 | 12 | 25% | $207,578 | 0.38% |
| 93. University of South Carolina | $52,400,000 | 25 | 0 | 0% | $50,181 | 0.10% |
| 94. University of New Hampshire | $52,360,000 | 9 | 0 | 0% | $19,105 | 0.04% |
| 95. Montana State University | $51,900,000 | 7 | 3 | 43% | $174,925 | 0.34% |
| 96. Arizona State University | $49,946,707 | 59 | 8 | 14% | $603,313 | 1.21% |
| 97. Mississippi State University | $47,712,143 | 21 | 8 | 38% | $169,958 | 0.36% |
| 98. University of Dayton | $45,621,626 | 16 | 2 | 13% | $753,396 | 1.65% |

| Name of Institution | FY 1997 Total Sponsored Research Expenditures | FY 1997 Invention Disclosures Received | FY 1997 Licenses & Options Executed | Percent of Invention Disclosures Licensed (%) | FY 1997 Gross License Income Received | Income as a Percentage of Research Expenditures |
|---|---|---|---|---|---|---|
| 99. Rice University | $44,800,000 | 12 | 3 | 25% | $36,000 | 0.08% |
| 100. Brandeis University | $43,786,511 | 23 | 6 | 26% | $135,967 | 0.31% |
| 101. University of Houston | $42,740,863 | 40 | 7 | 18% | $79,121 | 0.19% |
| 102. Louisiana State University | $41,842,198 | 25 | 3 | 12% | $413,226 | 0.99% |
| 103. North Dakota State University | $40,007,000 | 29 | 10 | 34% | $1,101,846 | 2.75% |
| 104. New Jersey Institute of Technology | $40,000,000 | 39 | 6 | 15% | $22,500 | 0.06% |
| 105. Idaho Research Fndtn./ Univ. of Idaho | $38,783,023 | 21 | 0 | 0% | $219,967 | 0.57% |
| 106. Kansas State University Research Fndtn. | $37,896,939 | 35 | 7 | 20% | $202,186 | 0.53% |
| 107. Syracuse University | $37,300,000 | 9 | 5 | 56% | $121,438 | 0.33% |
| 108. University of New Orleans | $37,229,856 | 4 | 1 | 25% | $32,159 | 0.09% |
| 109. University of Oregon | $36,609,701 | 8 | 3 | 38% | $183,000 | 0.50% |
| 110. Univ. of Oklahoma Health Science Center | $36,586,694 | 21 | 3 | 14% | $129,606 | 0.35% |
| 111. University of Maine | $33,819,599 | 6 | 1 | 17% | $0 | 0.00% |
| 112. Temple University | $33,000,000 | 34 | 10 | 29% | $826,177 | 2.50% |
| 113. University of Louisville | $29,494,000 | 20 | 6 | 30% | $91,320 | 0.31% |
| 114. University of Montana | $28,912,321 | 4 | 2 | 50% | $4,300 | 0.01% |
| 115. George Mason University | $27,037,000 | 8 | 3 | 38% | $8,861 | 0.03% |
| 116. Michigan Technological University | $26,500,000 | 25 | 11 | 44% | $166,846 | 0.63% |
| 117. New York Medical College | $23,866,247 | 10 | 5 | 50% | $108,891 | 0.46% |
| 118. Loyola University Medical Center | $23,400,000 | 8 | 7 | 88% | $225,000 | 0.96% |
| 119. Lehigh University | $23,242,767 | 6 | 3 | 50% | $105,887 | 0.46% |
| 120. Creighton University | $22,418,808 | 16 | 2 | 13% | $116,464 | 0.52% |
| 121. Southern Illinois Univ./ Carbondale | $20,549,063 | 14 | 4 | 29% | $250,624 | 1.22% |
| 122. Wright State University | $20,176,000 | 10 | 4 | 40% | $172,579 | 0.86% |

| Name of Institution | FY 1997 Total Sponsored Research Expenditures | FY 1997 Invention Disclosures Received | FY 1997 Licenses & Options Executed | Percent of Invention Disclosures Licensed (%) | FY 1997 Gross License Income Received | Income as a Percentage of Research Expenditures |
|---|---|---|---|---|---|---|
| 123. University of North Dakota | $18,849,495 | 1 | 0 | 0% | $0 | 0.00% |
| 124. Univ. of Maryland, Baltimore County | $18,155,000 | 18 | 1 | 6% | $29,094 | 0.16% |
| 125. Ohio University | $14,900,000 | 16 | 2 | 13% | $618,535 | 4.15% |
| 126. Brigham Young University | $12,795,126 | 38 | 14 | 37% | $2,365,563 | 18.49% |
| 127. Medical College of Ohio | $11,129,832 | 4 | 1 | 25% | $32,026 | 0.29% |
| 128. University of Akron | $10,274,943 | 15 | 2 | 13% | $149,444 | 1.45% |
| 129. Kent State University | $9,802,064 | 8 | 3 | 38% | $298,222 | 3.04% |
| 130. East Carolina University | $9,613,000 | 13 | 1 | 8% | $10,839 | 0.11% |
| 131. Portland State University | $9,099,037 | 3 | 0 | 0% | $0 | 0.00% |
| 132. University of Tulsa | $8,234,747 | 0 | 0 | NA | $0 | 0.00% |
|  |  |  |  | Avg.= 30.6% |  | Avg.= 2.17% |

**Total of 132 U.S. Universities**   $21,386,650,472   9,555   3,078   $576,889,538

NOTE: Data on Research Expenditures and Licensing Income for the universities listed were derived from AUTM's Licensing Survey FY 1998. Data are used with permission. For additional information please visit the Association of University Technology Managers, Inc. at www.autm.net. Also, please note that Percent of Invention Disclosures Licensed and Income as a Percent of Research Expenditures have been calculated by the book's authors.

# top 300 organizations receiving U.S. patents in 1998

APPENDIX

| Rank | Organization | Patents Received |
|------|--------------|------------------|
| 1 | IBM Corp. | 2657 |
| 2 | Canon K.K. | 1928 |
| 3 | NEC Corp. | 1627 |
| 4 | Motorola, Inc. | 1406 |
| 5 | Sony Corp. | 1316 |
| 6 | Samsung Electronics Co., Ltd | 1304 |
| 7 | Fujitsu, Ltd. | 1189 |
| 8 | Toshiba Corp. | 1170 |
| 9 | Eastman Kodak Co. | 1124 |
| 10 | Hitachi, Ltd. | 1094 |
| 11 | Mitsubishi Denki K.K. | 1080 |
| 12 | Matsushita Electric Industrial Co., Ltd. | 1034 |
| 13 | Lucent Technologies Inc. | 928 |
| 14 | Hewlett-Packard Co. | 805 |
| 15 | Xerox Corp. | 769 |
| 16 | General Electric Co. | 729 |
| 17 | U.S. Philips Corp. | 725 |
| 18 | Intel Corp. | 701 |
| 19 | Siemens A.G. | 626 |
| 20 | Texas Instruments, Inc. | 611 |

| Rank | Organization | Patents Received |
|------|--------------|------------------|
| 21 | Micron Technology, Inc. | 579 |
| 22 | Nikon Corp. | 575 |
| 23 | Sharp K.K. | 560 |
| 24 | Advanced Micro Devices, Inc. | 556 |
| 25 | Minnesota Mining and Manufacturing Co. | 554 |
| 26 | Fuji Photo Film Co., Ltd. | 547 |
| 27 | Ford Motor Co. | 461 |
| 28 | Procter and Gamble Co. | 454 |
| 29 | Sun Microsystems, Inc. | 423 |
| 30 | Ricoh Co., Ltd. | 406 |
| 31 | BASF A.G. | 399 |
| 32 | University of California | 395 |
| 33 | E. I. Du Pont de Nemours & Co. | 393 |
| 34 | Honda Motor Co., Ltd. | 389 |
| 35 | Toyota Jidosha K.K. | 387 |
| 36 | Bayer A.G. | 381 |
| 37 | Robert Bosch GmbH | 348 |
| 38 | Microsoft Corp. | 341 |
| 39 | Hoechst A.G. | 334 |
| 40 | Daewoo Electronics Co., Ltd. | 327 |
| 41 | Seiko Epson Corp. | 316 |
| 42 | Nippondenso Co., Ltd. | 312 |
| 43 | General Motors Corp. | 308 |
| 44 | Minolta Camera Co., Ltd. | 305 |
| 45 | Yazaki Corp. | 304 |
| 46 | Olympus Optical Co., Ltd. | 290 |
| 47 | Fuji Xerox Co., Ltd. | 278 |
| 48 | Asahi Kogaku Kogyo K.K. | 277 |
| 49 | Apple Computer, Inc. | 270 |
| 50 | Sanyo Electric Co., Ltd. | 256 |
| 51 | Compaq Computer Corp. | 245 |
| 52 | Medtronic, Inc. | 243 |
| 53 | Daimler-Benz A.G. | 241 |
| 54 | Lg Semicon Co., Ltd. | 234 |

| Rank | Organization | Patents Received |
|------|--------------|------------------|
| 55 | LSI Logic Corp. | 229 |
| 56 | Industrial Technology Research Institute, Taiwan | 218 |
| 57 | Taiwan Semiconductor Manufacturing Co., Ltd. | 218 |
| 58 | Murata Manufacturing Co., Ltd. | 217 |
| 59 | Whitaker Corp. | 216 |
| 60 | LG Electronics Inc. | 215 |
| 61 | Hyundai Electronics Industries Co., Ltd. | 212 |
| 62 | Ericsson, Inc. | 210 |
| 63 | Brother Kogyo K.K. | 209 |
| 64 | Telefonaktiebolaget L.M. Ericsson | 204 |
| 65 | Northern Telecom, Ltd. | 199 |
| 66 | Yamaha Corp. | 198 |
| 67 | Caterpillar Inc. | 197 |
| 68 | Northrop Grumman Corp. | 197 |
| 69 | L'Oreal S.A. | 182 |
| 70 | Eli Lilly and Co. | 180 |
| 71 | Merck & Co., Inc. | 178 |
| 72 | National Semiconductor Corp. | 178 |
| 73 | Applied Materials, Inc. | 176 |
| 74 | Dow Chemical Co. | 174 |
| 75 | SGS-Thomson Microelectronics, Inc. | 174 |
| 76 | United Microelectronics Corp. | 174 |
| 77 | NCR Corp. | 167 |
| 78 | Chrysler Motors Corp. | 166 |
| 79 | Konica Corp. | 166 |
| 80 | Abbott Laboratories | 164 |
| 81 | CIBA Specialty Chemicals Corp. | 164 |
| 82 | Digital Equipment Corp. | 164 |
| 83 | Eaton Corp. | 164 |
| 84 | Nissan Motor Co., Ltd. | 164 |
| 85 | Seagate Technology | 163 |
| 86 | Kimberly-Clark Worldwide, Inc. | 161 |
| 87 | Sumitomo Electric Industries Co., Ltd. | 158 |
| 88 | Becton, Dickinson and Co. | 157 |

| Rank | Organization | Patents Received |
|------|--------------|------------------|
| 89 | Boeing Co. | 157 |
| 90 | MITA Industrial Co., Ltd. | 154 |
| 91 | Henkel Corp. | 153 |
| 92 | AT&T Corp. | 150 |
| 93 | VLSI Technology, Inc. | 150 |
| 94 | Sumitomo Chemical Co., Ltd. | 148 |
| 95 | TRW Inc. | 144 |
| 96 | Shell Oil Co. | 143 |
| 97 | Hughes Electronic Devices Corp. | 142 |
| 98 | Institut Francias du Petrole | 142 |
| 99 | Agfa-Gevaert N.V. | 141 |
| 100 | Pioneer Electronic Corp. | 141 |
| 101 | Allied-Signal, Inc. | 140 |
| 102 | Bayer Corp. | 140 |
| 103 | Massachusetts Institute of Technology | 138 |
| 104 | Zeneca Ltd. | 138 |
| 105 | Bristol-Myers Squibb Co. | 137 |
| 106 | Genentech, Inc. | 133 |
| 107 | Sumitomo Wiring Systems, Ltd. | 132 |
| 108 | Heidelberger Druckmaschinen A.G. | 129 |
| 109 | KAO Corp. | 129 |
| 110 | Yamaha Motor Co., Ltd. | 129 |
| 111 | Cirrus Logic, Inc. | 128 |
| 112 | Semiconductor Energy Laboratory Co., Ltd. | 127 |
| 113 | Denso Corp. | 124 |
| 114 | Harris Corp. | 124 |
| 115 | Honeywell, Inc. | 123 |
| 116 | Novo Nordisk A/S | 122 |
| 117 | Unisys Corp. | 122 |
| 118 | Mitsubishi Chemical Corp. | 121 |
| 119 | Electronics and Telecommunications Research | 120 |
| 120 | Illinois Tool Works, Inc. | 120 |
| 121 | OKI Electric Industry Co., Ltd. | 120 |
| 122 | Pitney-Bowes, Inc. | 120 |

| Rank | Organization | Patents Received |
|------|--------------|------------------|
| 123 | Vanguard International Semiconductor Corp. | 120 |
| 124 | Exxon Chemicals Patents, Inc. | 119 |
| 125 | Philips Electronics North America Corp. | 118 |
| 126 | SGS-Thomson Microelectronics S.A. | 118 |
| 127 | MCI Communications Corp. | 117 |
| 128 | Pfizer, Inc. | 117 |
| 129 | Raytheon Co. | 117 |
| 130 | Aisin Seiki K.K. | 116 |
| 131 | United Technologies Corp. | 116 |
| 132 | American Cyanamid Co. | 112 |
| 133 | Rohm Co., Ltd. | 112 |
| 134 | McDonnell Douglas Corp. | 111 |
| 135 | Colgate-Palmolive Co. | 110 |
| 136 | Dell USA, L.P. | 109 |
| 137 | Dow Corning Corp. | 108 |
| 138 | Goodyear Tire and Rubber Co. | 108 |
| 139 | Lockheed Martin Corp. | 108 |
| 140 | Pioneer Hi-Bred International, Inc. | 108 |
| 141 | Hughes Electronics Corp. | 107 |
| 142 | Incyte Pharmaceuticals, Inc. | 107 |
| 143 | Akzo Nobel N.V. | 105 |
| 144 | Cypress Semiconductor Corp. | 105 |
| 145 | Exxon Research and Engineering Co. | 105 |
| 146 | Dai Nippon Printing Co., Ltd. | 103 |
| 147 | Fuji Photo Optical Co., Ltd. | 102 |
| 148 | Tokyo Electron Ltd. | 101 |
| 149 | Alcatel N.V. | 99 |
| 150 | Alps Electric Co., Ltd. | 99 |
| 151 | Merck Patent GmbH | 99 |
| 152 | Baker Hughes, Inc. | 97 |
| 153 | FMC Corp. | 97 |
| 154 | PPG Industries, Inc. | 97 |
| 155 | Baxter International, Inc. | 96 |
| 156 | Shin Etsu Chemical Co., Ltd. | 96 |

| Rank | Organization | Patents Received |
|------|-------------|:----------------:|
| 157 | University of Texas | 96 |
| 158 | BASF Corp. | 95 |
| 159 | Asea Brown Boveri A.G. | 94 |
| 160 | Delco Electronics Corp. | 94 |
| 161 | Nippon Steel Corp. | 94 |
| 162 | SmithKline Beecham Corp. | 94 |
| 163 | California Institute of Technology | 93 |
| 164 | Nokia Telecommunications OY | 93 |
| 165 | Hyundai Motor Co., Ltd. | 92 |
| 166 | NGK Insulators, Ltd. | 92 |
| 167 | United States Surgical Corp. | 92 |
| 168 | Silicon Graphics, Inc. | 90 |
| 169 | Southpac Trust International, Inc. | 89 |
| 170 | Commissariat A L'Energie Atomique | 88 |
| 171 | Nippon Telegraph & Telephone Corp. | 88 |
| 172 | UOP | 88 |
| 173 | Fuji Electric Co., Ltd. | 87 |
| 174 | Pacesetter, Inc. | 87 |
| 175 | Tetra Laval Holdings & Finance S.A. | 87 |
| 176 | Otis Elevator Company | 86 |
| 177 | Bridgestone Corp. | 85 |
| 178 | Samsung Display Devices Co., Ltd. | 84 |
| 179 | Morton International, Inc. | 83 |
| 180 | Sanshin Kogyo K.K. | 83 |
| 181 | TDK Corp. | 83 |
| 182 | W.L. Gore Associates, Inc. | 83 |
| 183 | Wisconsin Alumni Research Foundation | 83 |
| 184 | Dana Corp. | 82 |
| 185 | Qualcomm, Inc. | 82 |
| 186 | Chiron Corp. | 81 |
| 187 | SGS-Thompson Microelectronics, Inc. | 81 |
| 188 | Westinghouse Electric Corp. | 81 |
| 189 | Advantest Corp. | 80 |
| 190 | Air Products and Chemicals, Inc. | 80 |

| Rank | Organization | Patents Received |
|------|-------------|------------------|
| 191 | Boehringer Mannheim GmbH | 80 |
| 192 | Rhone-Poulenc Chimie | 80 |
| 193 | Molex, Inc. | 79 |
| 194 | Noritsu Koki Co., Ltd. | 79 |
| 195 | Stanford University, Leland Junior, Board of Trustees | 79 |
| 196 | Johns Hopkins University | 78 |
| 197 | Novartis Corp. | 78 |
| 198 | Schering Corp. | 78 |
| 199 | Toyoda Jidoshokki Seisakusho K.K. | 78 |
| 200 | Xilinx, Inc. | 78 |
| 201 | Aisin AW Co., Ltd. | 77 |
| 202 | Monsanto Co. | 77 |
| 203 | Rockwell International Corp. | 77 |
| 204 | Victor Company of Japan, Ltd. | 77 |
| 205 | Sandia Corp. | 76 |
| 206 | Sumitomo Rubber Industries, Ltd. | 76 |
| 207 | Ebara Corp. | 75 |
| 208 | Halliburton Energy Services | 75 |
| 209 | National Science Council | 75 |
| 210 | Phillips Petroleum Co. | 75 |
| 211 | Toyoda Gosei K.K. | 75 |
| 212 | Casio Computer Co., Ltd. | 74 |
| 213 | Fichtel and Sachs A.G. | 74 |
| 214 | Komatsu Ltd. | 74 |
| 215 | Black & Decker Inc. | 73 |
| 216 | British Telecommunication, PLC | 73 |
| 217 | Voith Sulzer Papiermaschinen GmbH | 73 |
| 218 | Corning, Inc. | 72 |
| 219 | Hoffmann-La Roche Inc. | 72 |
| 220 | Takeda Chemical Industries Ltd. | 72 |
| 221 | Agency of Industrial Science & Technology | 71 |
| 222 | Fanuc Ltd. | 71 |
| 223 | Rohm and Haas Co. | 71 |
| 224 | Thomson Consumer Electronics, Inc. | 71 |

| Rank | Organization | Patents Received |
|------|--------------|------------------|
| 225 | Allergan, Inc. | 70 |
| 226 | Deere and Co. | 70 |
| 227 | Analog Devices, Inc. | 69 |
| 228 | Trimble Navigation, Ltd. | 69 |
| 229 | University of Pennsylvania | 69 |
| 230 | Hoechst Celanese Corp. | 68 |
| 231 | Owens-Corning Fiberglass Technology, Inc. | 68 |
| 232 | Amoco Corp. | 67 |
| 233 | Mobil Oil Corp. | 67 |
| 234 | Nestec, S.A. | 67 |
| 235 | Integrated Device Technology, Inc. | 66 |
| 236 | Kobe Steel Ltd. | 66 |
| 237 | Mazda Motor Corp. | 66 |
| 238 | Tektronix, Inc. | 66 |
| 239 | Thomson-CSF | 66 |
| 240 | Bell Atlantic Network Services, Inc. | 65 |
| 241 | Carrier Corp. | 65 |
| 242 | Mitsubishi Jidosha Kogyo K.K. | 65 |
| 243 | Shin-Etsu Handotai Co., Ltd. | 65 |
| 244 | Asahi Glass Company, Ltd. | 64 |
| 245 | Cornell Research Foundation Inc. | 64 |
| 246 | Lever Bros. Co., Division of Conopco, Inc. | 64 |
| 247 | Praxair Technology, Inc. | 64 |
| 248 | Symbol Technologies, Inc. | 64 |
| 249 | Valmet Corp. | 64 |
| 250 | Aktiebolaget Astra | 63 |
| 251 | Cordis Corp. | 63 |
| 252 | Degussa A.G. | 63 |
| 253 | Eastman Chemical Corp. | 63 |
| 254 | France Telecom | 63 |
| 255 | Imperial Chemical Industries PLC | 63 |
| 256 | Unisia Jecs Corp. | 63 |
| 257 | Gas Research Institute | 62 |
| 258 | General Hospital Corp. | 62 |

| Rank | Organization | Patents Received |
|------|-------------|------------------|
| 259 | Shimano, Inc. | 62 |
| 260 | Union Carbide Corp. | 62 |
| 261 | Warner-Lambert Co. | 62 |
| 262 | Elf Atochem S.A. | 61 |
| 263 | Lear Corp. | 61 |
| 264 | Societe Nationale Industrielle Aerospatiale | 61 |
| 265 | United Techologies Automotive, Inc. | 61 |
| 266 | Daikin Industries Ltd. | 60 |
| 267 | G.D. Searle & Co. | 60 |
| 268 | NSK Ltd. | 60 |
| 269 | Polaroid Corp. | 60 |
| 270 | Quantum Corp. | 60 |
| 271 | Arco Chemical Technology, L.P. | 59 |
| 272 | EMC Corp. | 59 |
| 273 | Mannesmann A.G. | 59 |
| 274 | Michigan State University | 59 |
| 275 | Seiko Instruments, Inc. | 59 |
| 276 | Westinghouse Air Brake Co. | 59 |
| 277 | Winbond Electronics Corp. | 59 |
| 278 | Kawasaki Steel Corp. | 58 |
| 279 | Litton Systems, Inc. | 58 |
| 280 | Lucas Industries Public Ltd. Co. | 58 |
| 281 | Mitsubishi Jukogyo K.K. | 58 |
| 282 | Oracle Corp. | 58 |
| 283 | Avery Dennison Corp. | 57 |
| 284 | Bridgestone Sports Co., Ltd. | 57 |
| 285 | Emerson Electric Co. | 57 |
| 286 | Mitsui Petrochemical Industries Ltd. | 57 |
| 287 | Scripps Research Institute | 57 |
| 288 | Electronic Data Systems Corp. | 56 |
| 289 | Research Corporation Technologies, Inc. | 56 |
| 290 | W.R. Grace & Co. - Conn. | 56 |
| 291 | Columbia University | 55 |
| 292 | Ethicon, Inc. | 55 |

| Rank | Organization | Patents Received |
|------|--------------|:----------------:|
| 293 | Lexmark International, Inc. | 55 |
| 294 | Matsushita Electronics Corp. | 55 |
| 295 | Nokia Mobile Phones (U.K.) Ltd. | 55 |
| 296 | Daicel Chemical Industries Ltd. | 54 |
| 297 | M.A.N. Roland Druckmaschinen A.G. | 54 |
| 298 | Nokia Mobile Phones Ltd. | 54 |
| 299 | Valeo S.A. | 54 |
| 300 | Ajinomoto Co., Inc. | 53 |
|  | Aluminum Company of America | 53 |
|  | Berg Technology, Inc. | 53 |
|  | Daimler-Benz Aerospace Airbus GmbH | 53 |
|  | Huels A.G. | 53 |
|  | Iowa State University Research Foundation | 53 |
|  | Omron Corp. | 53 |
|  | Sci-Med Life Systems, Inc. | 53 |

NOTE: The Intellectual Property Owners Association compiled this list from data obtained from the U.S. Patent & Trademark Office. The numbers are patents issued during 1998 for which an assignment of title was recorded in the Office when the patent was issued. Patents in the names of subsidiaries, other related companies, or divisions normally are not combined with patents in the name of the parent. Agencies of the U.S. Government that received patents are not included.

# investment banks

APPENDIX

| | | | |
|---|---|---|---|
| A.G. Edwards & Sons, Inc. | St. Louis | MO | (314) 955-3000 |
| ABN AMRO, Inc. | Chicago | IL | (312) 855-7600 |
| ABN AMRO Rothschild | Chicago | IL | (312) 855-7600 |
| Adams Harkness & Hill, Inc. | Boston | MA | (617) 371-3900 |
| Advest Inc. | Hartford | CT | (860) 509-1000 |
| All -Tech Investment Group, Inc. | Montvale | NJ | (888) 328-8246 |
| Allen & Co., Inc. | New York | NY | (212) 832-8000 |
| Allen C. Ewing & Co. | Fernandina Beach | FL | (904) 354-5573 |
| American Fronteer Financial | Denver | CO | (303) 860-1700 |
| Anderson & Strudwick, Inc. | Richmond | VA | (800) 767-2424 |
| Ashtin Kelly & Co. | Naples | FL | (800) 770-8280 |
| | | | |
| Banc of America Securities LLC | San Francisco | CA | (415) 627-2000 |
| Banc Stock Financial Services, Inc. | Columbus | OH | (614) 848-3400 |
| BancBoston Robertson Stephens | Boston | MA | (617) 434-2200 |
| Barington Capital Group LP | New York | NY | (212 974-5700 |
| Barron Chase Securities, Inc. | Boca Raton | FL | (800) 747-1202 |
| Bear Stearns & Co., Inc. | New York | NY | (888) 473-3819 |
| Berthel Fisher & Co. Financial Services | Cedar Rapids | IA | (800) 356-5234 |
| Black and Co., Inc. | Portland | OR | (800) 325-0599 |

| | | | |
|---|---|---|---|
| BlueStone Capital Partners LP | New York | NY | (800) 236-0835 |
| Boenning & Scattergood, Inc. | Langhorne | PA | (215) 752-1551 |
| Brookstreet Securities Corp. | Irvine | CA | (949) 852-6800 |
| | | | |
| C.E. Unterberg Towbin | New York | NY | (212) 572-8000 |
| Cambridge Capital LLC | Convent Station | NJ | (973) 606-4444 |
| Capital Resources, Inc. | Washington | DC | (202) 466-5685 |
| Capital West Securities, Inc. | Oklahoma City | OK | (405) 235-5710 |
| Cardinal Capital Management, Inc. | Raleigh | NC | (919) 828-3900 |
| Charles Schwab & Co., Inc. | Phoenix | AZ | (602) 852-3500 |
| Charles Webb & Co. | Dublin | OH | (614) 766-8400 |
| Chase Securities, Inc. | New York | NY | (212) 270-6000 |
| CIBC Oppenheimer | New York | NY | (800) 999-6726 |
| Coast Partners Securities, Inc. | San Francisco | CA | (415) 781-1822 |
| Commonwealth Associates | New York | NY | (212) 829-5800 |
| Conning & Co. Hartford | Hartford | CT | (860) 520-1599 |
| | | | |
| D.A. Davidson & Co. | Great Falls | MT | (406) 727-4200 |
| Dain Rauscher Wessels | Minneapolis | MN | (612) 371-2800 |
| Daiwa Securities America, Inc. | New York | NY | (212) 612-7000 |
| Deutsche Bank Alex. Brown | Baltimore | MA | (410) 727-1700 |
| Deutsche Bank Securities | New York | NY | (212) 469-5000 |
| Dirks & Co., Inc. | New York | NY | (212) 832-6700 |
| DLJ Direct, Inc. | Jersey City | NJ | (800) 825-5723 |
| Donaldson Lufkin & Jenrette Securities | New York | NY | (212) 892-3000 |
| | | | |
| E*Offering | San Francisco | CA | (415) 618-6200 |
| E*Trade Securities | Palo Alto | CA | (650) 842-2500 |
| EBI Securities Corp. | Englewood | CO | (800) 289-5691 |

| | | | |
|---|---|---|---|
| FAC/Equities | Albany | NY | (800) 462-6242 |
| Fahnestock & Co., Inc | New York | NY | (212) 668-8000 |
| Ferris Baker Watts, Inc. | Baltimore | MD | (800) 436-2000 |
| First Albany Corp. | Boston | MA | (617) 228-3000 |
| First Analysis Securities Corp. | Chicago | IL | (312) 258-1400 |
| First Colonial Securities Group, Inc. | Marlon | NJ | (800) 828-7658 |
| First Equity Corp. | Catonsville | MD | (800) 922-5626 |
| First London Securities Corp. | Dallas | TX | (214) 220-0690 |
| First of Michigan | Grosse Point Farms | MI | (313) 882-9440 |
| First Security Van Kasper | San Francisco | CA | (800) 652-1747 |
| First Union Capital Markets Corp. | Charlotte | NC | (800) 275-3862 |
| Furman Selz | New York | NY | (212) 309-8200 |
| | | | |
| Gaines Berland, Inc. | Bethpage | NY | (516) 470-1000 |
| George K. Baum & Co. | Kansas City | MO | (816) 531-4900 |
| Gerard Klauer Mattison & Co., Inc. | New York | NY | (212) 885-4000 |
| Gilford Securities, Inc. | New York | NY | (212) 888-6400 |
| Global Financial Group, Inc. | Minneapolis | MN | (800) 321-1894 |
| Goldman Sachs & Co. | New York | NY | (212) 902-1000 |
| Gruntal & Co. LLC | New York | NY | (212) 820-8200 |
| | | | |
| H.C. Wainwright & Co., Inc. | Boston | MA | (800) 727-7176 |
| Hagerty Stewart | Newport Beach | CA | (949) 752-8766 |
| Hambrecht & Quist | San Francisco | CA | (415) 439-3000 |
| Hanifen Imhoff, Inc. | Denver | CO | (303) 296-2300 |
| HD Brous & Co., Inc. | Great Neck | NY | (800) 382-7170 |
| Hoak Breedlove Wesneski & Co. | Dallas | TX | (972) 960-4848 |
| Hoefer & Arnett, Inc. | San Francisco | CA | (415) 362-7111 |
| Hornblower & Weeks, Inc. | New York | NY | (212) 361-2266 |
| Howe Barnes Investments, Inc. | Chicago | IL | (800) 275-4693 |
| | | | |
| ING Baring Furman Selz LLC | New York | NY | (212) 409-7700 |
| Institutional Equity Corp. | Plainview | NY | (800) 426-7346 |

| | | | |
|---|---|---|---|
| Interstate/Johnson Lane Corp. | Charlotte | NC | (800) 929-0724 |
| Invemed Associates, Inc. | New York | NY | (212) 421-2500 |
| | | | |
| J.C. Bradford & Co. | Nashville | TN | (800) 251-1060 |
| J.J.B. Hilliard W.L. Lyons, Inc. | Louisville | KY | (800) 444-1854 |
| J.P. Morgan & Co. | New York | NY | (212) 483-2323 |
| J.P. Turner & Co. LLC | Atlanta | GA | (404) 479-8300 |
| J.W. Barclay & Co., Inc. | New York | NY | (800) 967-7002 |
| Janney Montgomery Scott Inc. | Philadelphia | PA | (215) 665-6000 |
| Janssen/Meyers Associates LP | New York | NY | (212) 742-4200 |
| Jefferies & Co., Inc. | Los Angeles | CA | (310) 445-1199 |
| John G. Kinnard and Co., Inc. | Minneapolis | MN | (800) 444-7884 |
| Joseph Charles & Associates, Inc. | Boca Raton | FL | (561) 391-9090 |
| Joseph Stevens & Co., Inc. | New York | NY | (800) 609-9000 |
| Josephthal & Co., Inc. | New York | NY | (800) 334-6696 |
| | | | |
| Kashner Davidson Securities Corp. | Sarasota | FL | (941) 951-2626 |
| Kaufman Bros. LP | New York | NY | (212) 292-8100 |
| Keefe Bruyette & Woods, Inc. | New York | NY | (212) 323-8300 |
| | | | |
| L. H. Friend Weinress Frankson & Presson | Irvine | CA | (949) 852-9911 |
| Ladenburg Thalmann & Co., Inc. | New York | NY | (212) 409-2000 |
| Laidlaw Global Securities, Inc. | New York | NY | (212) 376-8800 |
| Lazard Freres & Co. LLC | New York | NY | (212) 632-6000 |
| Legg Mason Wood Walker, Inc. | Baltimore | MD | (410) 539-3400 |
| Lehman Brothers | New York | NY | (212) 526-7000 |
| | | | |
| M.R. Beal & Co. | New York | NY | (212) 983-3930 |
| Macdonald Investments, Inc. | Fort Wayne | IN | (219) 423-6380 |
| Marion Bass Securities Corp. | Raleigh | NC | (800) 334-8172 |
| Mason Hill & Co. | New York | NY | (212) 425-3000 |
| May Davis Group, Inc. | New York | NY | (212) 775-7400 |
| McDonald & Co. Securities, Inc. | Cleveland | OH | (216) 443-2300 |

| | | | |
|---|---|---|---|
| Merrill Lynch & Co. | New York | NY | (212) 449-1000 |
| Millenium Financial Group | La Jolla | CA | (817) 427-9888 |
| Mills Financial Services, Inc. | Chicago | IL | (312) 977-3006 |
| Montgomery Securities | San Francisco | CA | (415) 981-8191 |
| Morgan Keegan & Co., Inc. | Memphis | TN | (800) 366-7426 |
| Morgan Stanley Dean Witter | New York | NY | (212) 392-2222 |
| | | | |
| National Securities Corp. | Seattle | WA | (206) 622-7200 |
| Needham & Co., Inc. | New York | NY | (212) 371-8300 |
| Neidiger Tucker Bruner ,Inc. | Denver | CO | (303) 825-1825 |
| Nesbitt Burns Securities, Inc. | New York | NY | (212) 355-7457 |
| Network 1 Financial Securities, Inc. | Red Bank | NJ | (800) 886-7007 |
| New York Broker, Inc. | Fairfax | VA | (703) 934-6065 |
| Nomura Securities International, Inc. | New York | NY | (212) 667-9300 |
| Nutmeg Securities Ltd. | Westport | CT | (203) 226-1857 |
| | | | |
| Oscar Gruss and Son, Inc. | New York | NY | (212) 759-2020 |
| | | | |
| Pacific Crest Securities, Inc. | Portland | OR | (503) 248-0721 |
| Pacific Growth Equities, Inc. | San Francisco | CA | (415) 274-6800 |
| PaineWebber, Inc. | New York | NY | (212) 713-2000 |
| Paradise Valley Securities, Inc. | Phoenix | AZ | (602) 953-7980 |
| Paribas Corp. | New York | NY | (212) 841-3000 |
| Pauli Johnson Capital & Research, Inc. | St. Louis | MO | (314) 863-3300 |
| Paulson Investment Co., Inc. | Portland | OR | (503) 243-6000 |
| Pennsylvania Merchant Group Ltd. | West Conshohocken | PA | (800) 366-3764 |
| Prime Charter Ltd. | New York | NY | (212) 977-0600 |
| Prudential Securities, Inc. | New York | NY | (212) 214-1000 |
| Punk Ziegel & Co. | New York | NY | (212) 308-9494 |
| Putnam Lovell de Guardiola & Thornton, Inc. | San Francisco | CA | (415) 283-1500 |
| | | | |
| Ragen Mackenzie, Inc. | Seattle | WA | (206) 343-5000 |
| Raymond James & Associates, Inc. | St. Petersburg | FL | (800) 248-8863 |

| | | | |
|---|---|---|---|
| Raymond James & Associates, Inc. | St. Petersburg | FL | (727) 573-3800 |
| RBC Dominion Securities Corp. | New York | NY | (212) 361-2700 |
| R.J. Steichen & Co | Minneapolis | MN | (800) 328-4836 |
| Robert W. Baird & Co., Inc. | Milwaukee | WI | (800) 792-2473 |
| Robinson-Humphrey Co. | Atlanta | GA | (404) 266-6000 |
| Rockcrest Securities LLC | Dallas | TX | (214) 599-0007 |
| Royce Investment Group, Inc. | Woodbury | NY | (516) 364-8300 |
| Ryan Beck & Co. | Livingston | NJ | (800) 342-2325 |
| Ryan Lee & Co., Inc. | McLean | VA | (703) 847-3100 |
| | | | |
| Salomon Smith Barney | New York | NY | (631) 851-1685 |
| Sandler O'Neill & Partners LP | Great Neck | NY | (516) 829-3410 |
| Sanford C. Bernstein & Co., Inc. | New York | NY | (212) 486-5800 |
| SBC Warburg Dillon Read, Inc. | New York | NY | (212) 943-0400 |
| Schneider Securities, Inc. | Denver | CO | (800) 822-0224 |
| Schroder & Co., Inc. | New York | NY | (212) 492-6000 |
| Scott & Stringfellow, Inc. | Richmond | VA | (804) 643-1811 |
| Seaboard Securities, Inc. | Florham Park | NJ | (973) 514-1500 |
| SG Cowen | San Francisco | CA | (415) 646-7200 |
| Simmons & Co. International | Houston | TX | (713) 236-9999 |
| SoundView Technology Group | Stamford | CT | (203) 462-7200 |
| Southeast Research Partners, Inc. | New York | NY | (212) 309-1540 |
| Southwest Securities | Dallas | TX | (214) 651-1800 |
| Spencer Edwards, Inc. | Englewood | CO | (303) 740-8448 |
| Stephens, Inc. | Little Rock | AR | (501) 374-4361 |
| Sterne Agee & Leach, Inc. | Birmingham | AL | (800) 633-4638 |
| Strasbourger Pearson Tulcin Wolff, Inc. | New York | NY | (212) 952-7500 |
| Suntrust Equitable Securities | Nashville | TN | (615) 780-9300 |
| Sutro & Co., Inc. | San Francisco | CA | (415) 445-8323 |

| | | | |
|---|---|---|---|
| TD Securities | New York | NY | (212) 827-7300 |
| Tejas Securities Group, Inc. | Austin | TX | (800) 846-6803 |
| Thomas Weisel Partners LLC | San Francisco | CA | (415) 364-2500 |
| Trident Securities, Inc. | Raleigh | NC | (800) 222-2618 |
| Tucker Anthony Cleary Gull | New York | NY | (212) 225-8000 |
| Tucker Anthony Cleary Gull Reiland & McDevitt, Inc. | Milwaukee | WI | (800) 221-2537 |
| Tucker Anthony, Inc. | Boston | MA | (617) 725-2000 |
| U.S. Bancorp Piper Jaffray, Inc. | Minneapolis | MN | (612) 342-6000 |
| UBS Securities | New York | NY | (212) 574-3000 |
| Vector Securities International, Inc. | Deerfield | IL | (847) 940-1970 |
| Vectormex, Inc. | New York | NY | (212) 407-5500 |
| Volpe Brown Whelan & Co. | San Francisco | CA | (415) 956-8120 |
| Vontobel Securities Ltd. | New York | NY | (212) 826-5600 |
| W.R. Hambrecht & Co. LLC | San Francisco | CA | (415) 551-8600 |
| Wachovia Securities, Inc. | Charlotte | NC | (704) 379-9000 |
| Walsh Manning Securities LLC | New York | NY | (516) 622-3100 |
| Warburg Dillon Read LLC | New York | NY | (212) 821-3000 |
| Wasserstein Perella Securities, Inc. | New York | NY | (212) 969-2700 |
| Weatherly Securities Corp. | New York | NY | (212) 412-7200 |
| Web Street Securities, Inc. | Deerfield | IL | (800) 932-8723 |
| Wedbush Morgan Securities | Los Angeles | CA | (213) 688-8000 |
| West America Securities Corp. | Westlake Village | CA | (800) 935-9378 |
| Westminster Securities Corp. | New York | NY | (800) 553-6428 |
| Westport Resources Investment Services | Westport | CT | (800) 935-0222 |
| Whale Securities Co. LP | New York | NY | (212) 484-2057 |
| William Blair & Co. | Chicago | IL | (312) 236-1600 |
| William R. Hough & Co. | St. Petersburg | FL | (727) 823-8100 |
| Wit Capital Corp. | New York | NY | (212) 253-4400 |

# government agencies involved in technology transfer

APPENDIX

## Access Technology Across Indiana (ATAIN)
Indstate.edu/ATAIN/

ATAIN is a not-for-profit corporation whose members are representatives of universities, government, hospitals and businesses in Indiana. ATAIN promotes economic development in Indiana through technology transfer. While emphasizing technology transfer, ATAIN promotes educational partnerships and collaboration. ATAIN's focus is on Indiana industry and consumer needs, with an eye toward building a technology base in today's evolving global economy. Through ATAIN, your company has access to Indiana's leading technology research institutions, facilities and experts.

## Air Force, U.S.
1501 Lee Highway
Arlington, VA 22209
Phone: 202-857-0717
Fax:    202-887-5093
www.af.mil/

The mission of the U.S. Air Force is to defend the United States through control and exploitation of air and space. Teamed with the Army, Navy and Marine Corps, the Air Force is prepared to fight and win any war if deterrence fails. To meet this challenge, the Air Force is extremely competent in air and space superiority, global attack, rapid global mobility, precision engagement, information superiority and agile combat support.

## Air Force Technology Transition Office

AF DUS&T Program
AFRL/XPTT
2310 8th Street
Building 167, Room 126
Wright Patterson AFB, OH 45433-7801
Phone: 937-656-9015
www.afrl.af.mil

The development of affordable and integrated technologies for the air and space forces is the responsibility of the Air Force Research Laboratory. For information your first stop should be the Technology Transition Office.

## Alabama Industrial Development Training

One Technology Court
Montgomery, AL 36116
Phone: 334-242-4158
Fax:    334-242-0299
TDD:    334-242-0298

The mission of Alabama Industrial Development Training (AIDT) is to provide quality workforce development for Alabama's new and existing industries and to expand the opportunities of its citizens through the jobs these industries create. AIDT, an institute of the state's Department of Postsecondary Education, encourages economic development through job-specific training. Training services are offered in many areas, ranging from welding to software engineering. AIDT's services are free of charge to new and expanding industries throughout the state.

## The Alaska Technology Transfer Center

430 W. Seventh Avenue, Suite 110
Anchorage, AK 99501
Phone: 907-274-7232
Fax:  907-274-9524

The Alaska Technology Transfer Center serves as a centralized technology transfer center focused on assisting Alaskan businesses in identification and application of technology to address commercial opportunities and problems in Alaska. They offer three main services—information, consultation and linkages. ATTC information services range from in-depth database searches for technical and scientific information to identification of experts to work on a specific project.

## Ames Laboratory

Iowa State University
311 Administrative Services Facility
Ames, IA 50011
Phone: 515-294-3483
Fax:  515-294-3751
www.external.ameslab.gov/

The Ames Laboratory focuses on basic research in materials and chemical sciences, engineering, metallurgy, physics and chemistry.

## Ames Research Center

Commercial Technology Office
Mail Code 202A-3
Moffett Field, CA 94035
Phone: 650-604-5000

The Ames Research Center is the designated Center of Excellence for Information Technology. The ARC conducts research activities, technology programs and flight projects that advance the nation's capabilities in civilian military aeronautics, space sciences and space applications. It has a wide variety of facilities for life, earth and space science research.

## Argonne National Laboratory
Industrial Technology Development Center
9700 South Cass Avenue
Building 900
Argonne, IL 60439
Phone: 630-252-2000
www.anl.gov/

The Argonne National Laboratory develops technologies that can be applied to assist companies in designing products, devising innovative industrial processes, substituting materials, and addressing environmental concerns. It has applied research for basic energy research in areas such as energy conservation, environmental science, fossil and nuclear fuels, biology and biomedicine and parallel computing architectures.

## Arkansas Center for Technology Transfer
West 20th Street
Fayetteville, AR 72701
Phone: 501-575-3747
http://actt.engr.uark.edu/contact.html

The mission of the University of Arkansas Center for Technology Transfer (ACTT) is to improve and strengthen the economy of the state by applying technology-based resources to foster the development and improvement of the industrial base within the state. ACTT recognizes that controlled industrial growth, either through increased productivity of existing industry or the introduction of new manufacturing facilities, has a positive influence on the economic health and quality of life for the people of Arkansas.

## Army, U.S.
Technology Transfer Office
Army Research Laboratory, AMSRL-CS-TT
2800 Powder Mill Road
Adelphi, MD 20783
Phone: 301-394-2410
www.arl.mil/tto

The Army laboratories and R&D centers have a wealth of technology, advanced facilities and expertise that can be used for more than national defense. Army technology can also help to produce a stronger civilian economy, but only in partnership with academia and U.S. industry, which can advance new technology and bring new products and processes to the marketplace.

## Army Domestic Technology Transfer Program
U.S. Army CECOM-RDED
AMSEL-RD-AS-TD
Ft. Monmouth, NJ 07703-5201
Phone: 732-532-4222
www.arl.mil/tto/ArmyDTT

The Army Domestic Technology Transfer Program seeks to promote the transfer of Army technology and expertise to the private sector, so that this technology can be exploited for improving U.S. competitiveness. One mechanism for such transfers is the use of cooperative R&D agreements (CRDAs) between Army R&D activities and U.S. industry.

## Army Research Laboratory Technology Transfer Programs
U.S. Army Research Laboratory
ATTN: AMSRL-CS-TT
2800 Powder Mill Road
Adelphi, MD 20783-1197
Phone: 301-394-2410
Fax:    301-394-5818
www.arl.mil/tto/

Not only must the Army Research Laboratory support the soldier in the field, it must now also use American defense technology in new ways to strengthen and expand our economy. It must focus on improving the competitiveness with the emerging regional economics—leading to greater investments in our infrastructure, more technological innovations and additional jobs for Americans. Five programs assist the ARL Technology Transfer Office to accomplish its mission at both the domestic and international level: the ARL-SBIR, ARL-International Tech Transfer, ARL Domestic Tech Transfer, Army Domestic Tech Transfer Program, and the Army Independent R&D Program.

## Association of Federal Technology Transfer Executives (AFT2E)
www.nttc.edu/aft2e.html

In 1992, a group of colleagues from a variety of federal laboratories began meeting to discuss the formation of a professional society. Out of those sessions the Association for Federal Technology Transfer Excellence (AFT2E) was created. The overall goal was to foster high standards of professionalism and ethics among persons engaged in the transfer of technology developed at federal laboratories.

## Brookhaven National Laboratory's
## Office of Technology Transfer

P.O. Box 5000
(Building 475D)
Upton, NY 11973
Phone: 516-344-7338
Fax: 516-344-3729
www.bnl.gov/TECHXFER/tech_transfer.html

The Brookhaven National Laboratory's programs are centered on areas such as basic energy sciences that emphasize fundamental research on the biological, chemical and physical phenomena underlying energy-related transfer, conversion and storage systems; life science and nuclear medicine research and medical applications of nuclear techniques; and high energy and nuclear physics.

## Center for Applied Information Technology (CAIT)

University of Dallas
1845 Irving
Irving, TX 75062
cait@caitud.org

The Center for Applied Information Technology is the community outreach arm of the Information Technology program at the University of Dallas (UD), Graduate School of Management. The purpose of CAIT is to enhance the research and application of information technology currently under way at the University. The center will aid in the transfer of technology and research developed at UD to the public and private sectors and to evaluate and determine the best way to apply these new technologies in their day-to-day activities.

## Defense Research & Engineering Office for Technology Transition

3030 Defense Pentagon
Washington, DC 20301
www.dtic.mil/techtransit/

## Department of Defense, U.S.

OASD(PA)/DPC
1400 Defense Pentagon, Room lE757
Washington, DC 20301-1400
Phone: 202-563-3467
www.defenselink.mil/

The Department of Defense (DoD) is responsible for providing the military forces needed to deter war and protect the security of our country. It is responsible for providing information to military members, DoD civilians, military family members, the American public, Congress and the news media.

## Department of Defense Cooperative Programs for Reinvestment

4201 Wilson Boulevard, Suite 585
Arlington, VA 22230
Phone: 703-306-1380
Fax:  703-306-0290
www.nemonline.org/trp.html

The Technology Reinvestment Project's (TRP) goal is to develop militarily useful, commercially viable technology in order to improve the U.S. Department of Defense's access to affordable, advanced technology.

## Department of Defense Office of Technology Transition
1400 Defense Pentagon
Washington, DC 20585
Phone: 202-563-3467
www.dtic.mil/

The Department of Defense Office of Technology Transition was created to monitor all R&D activities by or for the military departments and defense agencies and to identify all R&D that use technologies or technological advancements having potential non-defense commercial applications. It coordinates and facilitates the transition of such technologies from the DoD to the private sector, and provides private firms with assistance involved in the transition of this technology.

## DoD SBIR / STTR / Fast Track
Phone: 800-382-4634
Fax:  800-462-4128
E-mail: SBIRHELP@teltech.com
www.acq.osd.mil/sadbu/sbir/

The DoD's SBIR program funds early-stage R&D projects at small technology companies—projects which serve a DoD need and have the potential for commercialization in private sector and/or military markets. The DoD issues an SBIR solicitation twice a year, describing its R&D needs and inviting R&D proposals from small companies organized for profit with 500 or fewer employees.

The DoD's STTR is similar in structure to SBIR, but funds cooperative R&D projects involving a small business and a research institution (i.e., university, federally funded R&D center, or nonprofit research institution). The purpose of STTR is to create, for the first time, an effective vehicle for moving ideas from our nation's research institutions to the market, where they can benefit both private sector and military customers.

The DoD's SBIR and STTR programs have featured a "Fast Track" process for SBIR/STTR projects that attract outside investors who will match phase II funding, in cash, contingent on the project's selections for phase II award. It provides a significantly higher chance of SBIR/STTR award, and continuous funding, to small companies that can attract outside investors.

(Small companies retain the patent rights to any inventions they develop under these programs. Funding is awarded competitively, but the process is streamlined and user-friendly.)

### Department of Energy, U.S.
The National Energy Information Center
1000 Independence Avenue, S.W.
Room 1F-048
Washington, DC 20585
Phone: 202-727-1800
Fax: 202-586-0727
www.eia.doc.gov/

The Department of Energy (DOE) (established in 1977) brought loosely defined government programs together to maximize efficiency. The DOE looks at every aspect of energy consumption and production and the efficiency of energy for transportation and within buildings. Its laboratories conduct research on materials, biotechnology, manufacturing, communications, aerospace, transportation, pollution minimization and remediation and energy technology.

### DOE Technology Information Network
1000 Independence Avenue, SW
Washington, DC 20585
Phone: 202-727-1800
www.dtin.doe.gov/

The DOE Technology Information Network (DTIN) is designed to be used as a resource for finding partnership opportunities through technology transfer available at the DOE facilities. The DTIN is a resource network for identifying DOE facilities' scientific and technical capabilities, equipment available for usage, technical staff expertise and licensable products.

## Dryden Flight Research Center
Technology Office
P.O. Box 273
Edwards, CA 93523
Phone: 661-258-3802
Fax:  661-258-3088

The Dryden Flight Research Center conducts safe and timely flight research for discovery, technology development and technology transfer for U.S. aeronautics and space preeminence.

## Ernest Orlando Lawrence Berkeley National Laboratory
University of California
Technology Transfer Department
1 Cyclotron Road, Bldg. 50A/4112
Berkeley, CA 94720
Phone: 510-486-6467
Fax:  510-486-6457
www.lbl.gov/Tech-Transfer/

The Lawrence Berkeley Laboratory is one of three Department of Energy Centers for research on the human genome. It conducts research in the areas of general and biological sciences and energy. The lab also has programs covering molecular genetics, medical imaging, carcinogenesis, gene expression, biomedical research, energy efficiency in buildings and advanced materials development.

**Far West Regional Technology Transfer Center (FWRTTC)**
3716 S. Hope Street, Suite 200
Los Angeles, CA 90007
Phone: 800-642-2872
Fax: 213-746-9043
www.usc.edu/dept/engineering/TTC

The FWRTTC is located at the University of Southern California. It has a technical staff of engineers and scientists highly qualified in their fields to help define and clarify problems and analyze ideas. The FWRTTC has a staff to identify opportunities at specific federal labs and other useful resources.

**Goddard Space Flight Center**
Office of Commercial Programs
Mail Code 702, Greenbelt Road
Greenbelt, MD 20771
Phone: 301-286-2000 / 286-5810
Fax: 301-286-0301

The mission of the Goddard Space Flight Center is to expand knowledge of the Earth and its environment, the solar system and the universe through observations from space. To assure that our nation maintains leadership in this endeavor, the center is committed to excellence in scientific investigation, in the development and operation of space systems and in the advancement of essential technologies.

## Hawaii's High Technology Development Corporation

2800 Woodlawn Drive, Suite 100
Honolulu, HI 96822
Phone: 808-539-3806
Fax:  808-539-3611
www.htdc.org/index

The High Technology Development Corporation (HTDC) is a state agency in Hawaii that was developed to promote the growth of commercial high technology, which develops industrial parks and supports programs for high-tech companies. The HTDC also advises the governor on the formulation of technology-based economic development policy.

## Hawaii's Manoa Innovation Center

2800 Woodlawn Drive, Suite 100
Honolulu, HI 96822
Phone: 808-539-3600
Fax:  808-539-3614

The Manoa Innovation Center helps to unite emerging commercial ventures with university-oriented applied research organizations.

## Hawaii's Maui Research & Technology Center

590 Lipoa Parkway
Kihei, Maui, HI 96753
Phone: 808-875-2320
Fax:  808-875-2329

The Maui Research and Technology Center (MRTC) is a combination of a research facility and an incubator facility for start-up companies at the University of Hawaii, Maui. It also offers a facility for high-tech companies that are relocating to Maui. The MRTC has a state-of-the-art telecommunications and computing center, which offers access to a federally funded Imaging Information Center also located on Maui.

## Hawaii Small Business Innovation Research Grant Program
High Technology Development Corporation
2800 Woodlawn Drive, Suite 100
Honolulu, HI 96822
Phone: 808-539-3600
Fax:  808-539-3611
mana.htdc.org/sbir/sbir.html

The Hawaii Small Business Innovation Research (SBIR) Grant Program was established to encourage Hawaii companies to participate in the federal SBIR Program, to enhance competitiveness of Hawaii companies for Phase II awards, and to increase the potential for full commercial development of Hawaii innovation.

## Idaho National Engineering Laboratory
Technology Transfer Management
P.O. Box 1625
Idaho Falls, ID 83415
Phone: 208-526-4430
www.inel.gov/technology_transfer/techtran.html

The Idaho National Engineering Laboratory has scientists specializing in materials and material processing, information and communications technology, robotics, mechanical and electronic system development, computational intelligence, environmental and waste treatment technology, engineering and chemical sciences, biotechnology, nuclear reactor research technology and instrumentation development.

## Jet Propulsion Laboratory's Technology Transfer Page
Technology Transfer Office
4800 Oak Grove Drive
Pasadena, CA 91109
Phone: 818-354-2577
Fax:  818-393-2754
www.jpl.nasa.gov

Managed for NASA by the California Institute of Technology, the Jet Propulsion Laboratory (JPL) is the leading U.S. center for robotic exploration of the solar system. In addition to working for NASA, JPL performs research for a number of federal agencies.

## Johnson Space Center
Office of Technology Transfer
2101 NASA Road 1, MC-HA
Houston, TX 77058
Phone: 281-483-3809
www.jsc.nasa.gov/

The Johnson Space Center was established as NASA's primary center for design and development testing of spacecraft and associated systems for human flight; for selection and training of astronauts; for planning and conducting human space flight missions and for extensive participation in the medical and engineering scientific experiments carried aboard space flights.

## Kennedy Space Center Technology Transfer Office
Phone: 407-867-5000
technology.ksc.nasa.gov/

The activities and opportunities of the Technology Transfer Program at Kennedy Space Center are providing new technology and problem solutions to industries located throughout the country. The transfer of this new technology serves to strengthen the nation's economy and benefit the general public.

## Langley's Technology Applications Group

11 Langley Boulevard
Hampton, VA 23681
Phone: 757-864-6005
Fax:  757-864-8088

The mission of the NASA Langley Research Center is to increase the knowledge and capability of the United States in a full range of aeronautics disciplines and in selected space disciplines. The Technology Applications Group is responsible for introducing, facilitating, promoting and supporting technology transfer and commercialization of advanced aeronautics, space and related technologies.

## Lawrence Livermore National Laboratory

Industrial Partnerships and Commercialization
P.O. Box 808, L-795
Livermore, CA 94550
Phone: 925-423-5660
Fax:  925-423-8988
www.llnl.gov/

The Lawrence Livermore National Laboratory's research addresses the situation of nuclear danger. Its defense-related work has been redirected to implement Start I and II treaties.

## Lewis Research Center

Office of Commercial Technology Office
21000 Brookpark Road
Cleveland, OH 44135
Phone: 216-433-3484
Fax:  216-433-5531
www.grc.nasa.gov

Lewis is NASA's lead center for research, technology and development in aircraft propulsion, space propulsion, space power and satellite communications. The center has been advancing propulsion technology to enable aircraft to fly faster, farther and higher, and also has focused its research on fuel economy, noise abatement, reliability and reduced pollution. The center pioneered efforts in the use of high-energy fuels for both air breathing and space propulsion. Projects demonstrated the practicality of liquid hydrogen as a fuel, leading to its use in the Apollo and the Space Shuttle programs as prime examples.

## Los Alamos National Laboratory
Industrial Partnership Office
2237 Trinity Drive, MS C331
Los Alamos, NM 87545
Phone: 505-665-9090
Fax:  505-665-0154
www.lanl.gov/Internal/projects/IPO/welcome.html

The mission of Los Alamos National Laboratory is enhancing the security of nuclear weapons and nuclear materials worldwide. As well as focusing on nuclear safeguards and security, LANL has programs on energy, biomedical science, high-performance computing research, materials science and human genome studies.

## Marshall Space Flight Center Technology Transfer Office
Mail Code LA01
Marshall Space Flight Center
Huntsville, AL 35812
Phone: 256-544-4266
Fax:  256-544-1815
www.msfc.nasa.gov/

Marshall Space Flight Center is primarily organized around four Directorates. These Directorates are Space Transportation, Science, Flight Projects and Engineering. The Space Transportation Directorate is designed to implement Marshall's role as NASA's lead center for space transportation development. The objective of this

Directorate is to make space transportation more accessible, reliable, safer and less costly. Responsibilities for the Science Directorate include original research and technology development in the areas of low-gravity materials science and biotechnology, earth and space science and space optics manufacturing. The international space station is the primary focus of the Flight Projects Directorate. This Directorate manages Marshall's assigned payload and mission integration and operations, as well as payload utilization carrier integration and life support systems for air and water. In addition to the space station activities, the Flight Projects Directorate oversees the development of space solar power.

## Mid-Atlantic Technology Applications Center

3400 Forbes Avenue
Pittsburgh, PA 15260
Phone: 412-383-2500
Fax: 412-383-2595
lhummel@mtac.pitt.edu

The Mid-Atlantic Technology Applications Center (MTAC) was established as part of NASA's commitment to broaden the scope and increase the effectiveness of its technology commercialization program. As one of NASA's six Regional Technology Transfer Centers, MTAC helps U.S. firms improve their competitiveness by assisting them in the location, assessment, acquisition and utilization of technologies and scientific and engineering expertise within the federal government. In pursuit of this mandate, MTAC developed an in-depth knowledge of industry needs and, as a result, now offers a technology marketing service for federal, university and corporate laboratories.

## Mid-Continent Technology Transfer Center (MCTCC)

301 Tarrow
College Station, TX 77843-8000
Phone: 409-845-2907
Fax: 409-845-3559

As one of six NASA-funded Regional Technology Transfer Centers, MCTTC offers a variety of technology transfer and commercialization services. It serves private companies and federal laboratories by forging a value-added link between technology sources and recipients. It provides this assistance to a 14-state region from its headquarters in College Station, Texas, and through a team of subcontractors and affiliates. The team, composed of universities, private industry, federal and state agencies, caters primarily to companies in the mid-continent region, but it pulls essential technologies from more than 600 laboratories nationwide.

## Midwest (Great Lakes) Technology Center (GLITeC)

25000 Great Northern Corporate Center, Suite 260
Cleveland, OH 44070
Phone: 440-734-0094
Fax: 440-734-0686
www.battelle.org/GLITeC

Through targeted technology management services, GLITeC works with industry to acquire and use NASA technology and expertise. GLITeC has dedicated in-reach into NASA and bridges the gap between federal discovery and commercial application. GLITeC and its regional affiliates provide industry with tools and methodology to develop new products and enhance processes through the efficient application of technology and related capabilities from NASA. Its services are designed to help companies identify, acquire, adapt and utilize the federal technology and capabilities.

## National Aeronautics and Space Administration (NASA) Commercial Technology Network (CTN)
www.nctn.hq.nasa.gov

The NASA CTN is a network that was formed to assist in providing information to help transfer technology from NASA's ten laboratories and external agents directly to industry. By connecting all the field centers and headquarters, they more efficiently serve NASA and the customers needs.

## NASA Regional Technology Transfer Centers
www.nctn.hq.nasa.gov

NASA-sponsored Regional Technology Transfer Centers expedite technology transfer and spur economic development. The program divides the nation into six regions and is implemented through regional networks that provide organizations with local access to direct and timely services. The RTTCs have professional staffs with extensive business and industry experience that work with companies to locate, access, acquire and use technologies and expertise from federal laboratories, state programs and private industry. RTTCs also help U.S. companies find and use technologies, expertise and facilities at NASA and more than 1,000 other federal, state, university and industry laboratories.

## NASA Small Business Innovation Research Program
Office of Director, Commercial Development
and Technology Transfer
NASA Headquarters
300 E Street, SW
Washington, DC 20546
Phone: 202-358-2320
Fax:  202-358-3878

## National Cancer Institute (NCI)

Inquiries Office
Building 31, Room 10A03
31 Center Drive
Bethesda, MD 20892-2580
Phone: 301-435-3848
Fax:  301-402-0894
www.nci.nhi.gov

The National Cancer Institute is a component of the National Institutes of Health, one of eight agencies that compose the Public Health Service in the Department of Health and Human Services. The National Cancer Institute coordinates the National Cancer Program, which conducts and supports research, training, education, health information dissemination and other programs with respect to the cause, diagnosis, prevention and treatment of cancer, rehabilitation from cancer and the continuing care of cancer patients and the families of cancer patients.

## NCI's Technology Transfer Fellowship Program

31 Center Drive
Bethesda, MD 20892-2580
Phone: 301-435-3848
www.nci.nih.gov/TTRAN/TTFP/TTF

The Technology Transfer Fellowship Program sponsored by the National Cancer Institute provides fellowships to individuals who hold advanced degrees in a number of fields, including science, medicine, law and communications. Applicants don't need a background in technology transfer, but should have an interest in the application of technology transfer to cancer research and treatment. Fellowships are for 1 to 2 years, with possible renewal for up to 5 years.

## National Institute of Allergy and Infectious Diseases
Office of Communications and Public Liaison
Building 31, Room 7A-50
31 Center Drive MSC 2520
Bethesda, MD 20892-2520
Phone: 301-496-4000
www.niaid.nih.gov/director/director.htm

The National Institute of Allergy and Infectious Diseases (NIAID) had its origins in the earliest days of the Public Health Service. Today, NIAID provides the major support for scientists conducting research aimed at developing better ways to diagnose, treat and prevent the many infectious, immunologic and allergic diseases that afflict people worldwide.

## National Institute of Standards and Technology
Technology Services
Quince Orchard & Clopper Roads
306 NIST North
Gaithersburg, MD 20899
Phone: 301-975-4500
Fax:  301-975-2183
www.ts.nist.gov/ts

The National Institute of Standards and Technology (NIST) is a great resource for finding information about new technologies. It is the main research and development laboratory for the federal government. NIST strengthens the U.S. economy and improves the quality of life by working with industry to develop and apply technology, measurements and standards and to facilitate rapid commercialization of products based on new scientific discoveries.

## National Technology Transfer Center

316 Washington Avenue
Wheeling, WV 26003
Phone: 800-678-6882
Fax:  304-243-4388
technology@nttc.edu
www.nttc.edu

The National Technology Transfer Center (NTTC) was established in 1989 by Congress to provide companies and individuals with access to federal R&D to better enable them to compete in the international marketplace. Staffed with technology transfer specialists, technical area experts and information management professionals, the Center provides technology transfer support via several different areas: access to $70 billion worth of research and development, 100,000 research professionals at over 700 federal laboratories and universities; technology assessment services; product testing and prototyping and professional training and development. These services build upon each other to create NTTC's full service commercialization center.

## Naval Surface Warfare Center

Crane Division
300 Highway 361
Crane, IN 47522-5183
Phone: 703-602-0644
www.hq.dt.navy.mil/

The mission of the Naval Surface Warfare Center is to provide quality and responsive engineering and industrial base support of weapons systems, subsystems, equipments and components as assigned by the Commander, Naval Surface Warfare Center.

## NIH (National Institutes of Health)
Office of the Director
Building 1, Room B1-60
Bethesda, MD 20892
Phone: 301-496-4000
www.nih.gov

The NIH mission is to uncover new knowledge that will lead to better health for everyone. NIH works toward that mission by conducting research in its own laboratories; supporting the research of non-federal scientists in universities, medical schools, hospitals and research institutions throughout the country and abroad; helping in the training of research investigators and fostering communication of biomedical information.

## NIH's Interagency Edison
6701 Rockledge Dr., Rm. 3190, MSC 7750
Bethesda, MD 20892-7750
Phone: 301-435-0679
Fax:  301-480-0272
www.iedison.gov

Interagency Edison supports a "Common Face" for Invention Reporting to the Government. The system has been designed to facilitate grantee/contractor institutions' compliance with the laws and regulations mandated by the Bayh-Dole Act whose purpose is to ensure transfer of technology from the research laboratory to the commercial/public sector. Interagency Edison provides federal grantee/contractor organizations and participating federal agencies with the technology to electronically manage extramural invention portfolios in compliance with federal reporting requirements.

## NIH Office of Technology Transfer

Office of Technology Transfer
National Institutes of Health
6011 Executive Blvd., Suite 325
Rockville, MD 20852
Phone: 301-496-7057
Fax: 301-402-0220
www.nih.gov/od/ott/director.htm

The NIH Office of Technology Transfer (OTT) evaluates, protects, monitors and manages the NIH invention portfolio to carry out the mandates of the FTTA (Federal Technology Transfer Act 1986). This is largely accomplished through overseeing patent prosecution, negotiating and monitoring licensing agreements and providing oversight and central policy review of Cooperative Research and Development Agreements. OTT also manages the patent and licensing activities for the Food and Drug Administration (FDA). OTT is responsible for the central development and implementation of technology transfer policies for four research components of the PHS—the NIH, the FDA, the Centers for Disease Control and Prevention and the Agency for Health Care Policy and Research.

## North Carolina's Technology Transfer Center for Local Transportation Agencies

North Carolina State University
Centennial Campus Box 8601
Raleigh, NC 27695-8601
Phone: 919-515-8899
Fax: 919-515-8898
itre.ncsu.edu/LTAP/

The North Carolina Technology Transfer Center is one of 59 Local Technical Assistance Program centers nationwide. The objective of the center is to meet the needs of local transportation personnel agencies by establishing and maintaining a system to increase transportation expertise to the state and local transportation and public works agencies; to provide a conduit for materials prepared at the national level for local use; to encourage and promote effective implementation and use of research findings and innovations for improving transportation and to improve communication on transportation technology between federal, state and local transportation agencies.

## Northeast Regional Technology Transfer Center

1400 Computer Road
Westboro, MA 01581
Phone: 508-870-0042
Fax:  508-366-0101

The Center of Technology Commercialization (CTC), a non-profit company based in Westboro, MA, is NASA's Northeast Regional Technology Transfer Center, covering the six New England States plus New York and New Jersey. CTC has seven satellite offices (two in New York, with Vermont and New Hampshire covered from the Nashua, NH office). Acting as a gateway for the transfer of NASA and other federal technology to private industry, CTC is one of six RTTCs providing a nationwide network devoted to the common mission of assisting American industry to improve its worldwide competitiveness.

## Oak Ridge National Laboratory

Office of Technology Transfer
P.O. Box 2009
Oak Ridge, TN 37831
Phone: 423-576-8368
Fax:  423-576-9465
www.ornl.gov/patent/ornl_ott.html

The Oak Ridge National Laboratory's Office of Technology Transfer enables private industry and academia to make practical application of the advanced R&D and technical expertise at Oak Ridge. The program makes available the most innovative equipment, the latest technology and the valuable expertise of its staff to outside organizations. Hundreds of technologies have been developed and patented at the government-owned research and production facilities in Oak Ridge.

## Office of Air Quality Planning and Standards Technology Transfer Network
TTNBBS.RTPNC.EPA.GOV

The Office of Air Quality Planning and Standards Technology Transfer is a network of independent Bulletin Board Systems that provide technical information, documents, files and messages related to the control of air pollution.

## Office of Technology Innovation (OTI)
Office of Industrial Technologies
1000 Independence Avenue, SW
Washington, DC 20585-0121
Phone: 202-586-9232
Fax:  202-586-9234

The Office of Industrial Technologies (OIT) develops and delivers advanced energy efficiency, renewable energy and pollution prevention technologies for application in the U.S. industrial sector. OIT partners with industry, government and non-governmental organizations, with the goal of significantly improving the resource efficiency and competitiveness of materials and process industries. OIT is part of the Department of Energy's Office of Energy Efficiency and Renewable Energy.

## Pacific Northwest National Laboratory
P.O. Box 999
902 Batelle Boulevard
Richland, WA 99352
Phone: 509-375-2121
Toll Free: 888-375-7665
Fax:  509-372-4791
E-mail: inquiry@pnl.gov
www.pnl.gov

The Pacific Northwest Laboratory's research and development is focused on meeting national needs related to the economy, energy, the environment and the nation's security. Several of the laboratory's activities deal with global climate change and environmental issues.

## Phillips Laboratory Technology Transfer Office
3550 Aberdeen Avenue, SE
Kirtland Air Force Base, NM 87117
Phone: 505-846-6952
Fax:  505-846-5034

The Phillips Laboratory Technology Transfer Office promotes new space and missile technologies, propulsion, advanced weapons and survivability, lasers and imaging, space experiments and geophysics.

## Rhode Island Technology Transfer Center
580 Ten Rod Road
North Kingstown, RI 02852
Phone: 401-294-4400
Fax:  401-294-4220
www.rittc.com

The Rhode Island Technology Transfer Center is an ARPA- and NASA-sponsored organization that facilitates the transfer of technologies developed in federal laboratories to the commercial market.

**Rome Laboratory**
26 Electronic Parkway
Rome, NY 13441
Phone: 315-330-1860
Fax:  315-330-3022

The Rome Laboratory specializes in information displays, communications gear, radar, computer and intelligence systems.

**Sandia National Laboratories**
Technology Partnerships & Commercialization
P.O. Box 5800, Mail Stop 1380
Albuquerque, NM 87185
Phone: 505-843-4164
www.sandia.gov/

As a Department of Energy national laboratory, Sandia works in partnership with universities and industry to enhance the security, prosperity and well being of the nation. It provides scientific and engineering solutions to meet national needs in nuclear weapons and related defense systems, energy security and environmental integrity and to address emerging national challenges for both government and industry.

**Savannah River Site**
P.O. Box A
Aiken, SC 29802
Phone: 803-725-6211
Fax:  803-725-4657
www.srs.gov/

The mission of the Savannah River Site is to serve the national interest by ensuring that programs, operations and resources are managed in a safe, open and cost-effective manner to support current and future national security requirements; reducing global nuclear proliferation danger; protecting and restoring the environment while managing waste and nuclear materials and conducting mission-supportive research and technology development.

## Small Business Innovation Research Program

Norman Taylor
100 Bureau Drive, Stop 2000
Gaithersburg, MD 20899-2000
Phone: 301-975-4517
Norman.Taylor@nist.gov

The Small Business Innovation Research Program provides funding on a competitive basis to small high-tech businesses that can carry out research on topics of interest identified by NIST's Measurement and Standards Laboratories. Annually in October, NIST issues a list of research and development topics for which proposals are solicited. There are two phases of awards. In Phase I, awardees can receive up to $75,000 for a 6-month study to establish the technical feasibility of a proposed project. Successful Phase I participants may compete in Phase II for up to $300,000 to support further development of the work for a period not to exceed 2 years.

## Southern Technology Applications Center

1900 SW 34th Street, Suite 206
Gainesville, FL 32608-1260
Phone: 352-294-7822
Fax: 352-294-7802
STACINF@nervm.nerdc.ufl.edu

The Southern Technology Applications Center is one of six Regional Technology Transfer Centers that can help organizations locate and access technologies, expertise and capabilities within the federal laboratory system and the region's universities to solve technology-related problems and create commercial opportunities.

## Technical Opportunities Showcase

TOPS Implementation Office
MS 213
NASA Langley Research Center
Hampton, VA 23681-0001
Phone: 757-864-5800

NASA Langley's Technology Opportunities Showcase is a three-day exposition to facilitate the transfer of technology from government research to U.S. industry. The primary goal of the Showcase is to develop partnerships with aerospace and non-aerospace companies for expediting technology transfer and, thereby, bolstering the U.S. economy.

## Technology Transfer Information Center (Agricultural Research Service, USDA)

National Agricultural Library
10301 Baltimore Avenue, 4th Floor
Beltsville, MD 20705
Phone: 301-504-6875
Fax:  301-504-7098
ttic@nal.usda.gov

The Technology Transfer Information Center helps to promote the rapid conversion of federally developed inventions into commercial products by "getting the results of research into the hands of those individuals and organizations who can put it into practical use." It provides a variety of services to professionals involved in the innovation process by providing information support systems for the Agricultural Research Service, transferring technologies for industry, developing reference collections and facilitating information exchange.

## Technology Transfer Office NAWCWPNS—China Lake
Technology Development Projects Office
Code 52DA003
Point Mugu, CA 93042
Phone: 805-989-9208
Fax: 805-989-7855
www.happy.mugu.navy.mil/tt/mission.html

The NAWCWPNS's mission is to be a full-spectrum research and development, test and evaluation and in-service engineering center for weapons systems associated with air warfare, missiles and missile subsystems; aircraft weapons integration and airborne electronics warfare systems and to maintain and operate an air, land and sea range complex. Point Mugu is the Navy's premier test and evaluation center for Navy weapons.

## Technology Transfer Office of Naval Research Laboratory
Department of the Navy
4555 Overlook Avenue, SW
Washington, DC 20375
Phone: 202-767-3744
Fax: 202-404-7920
Infoweb.nrl.navy.mil/~techtran/

The Naval Research Laboratory (NRL) is responsible for scientific research and technological development in the areas of ocean, space, atmospheric and materials science and technology, as well as in the areas of warfare systems and sensors research.

**Technology Transfer Programs at Stennis Space Center**
Technology Transfer Office
Commercial Technology Program
NASA Stennis Space Center
Mail Code TA00
Stennis Space Center, MS 39529
Phone: 800-688-1929
Fax:  228-688-2408
technology@ssc.nasa.gov

The Small Business Technology Transfer Program at Stennis Space Center awards contracts to small business concerns for cooperative research and development with a research institution. Though modeled after the Small Business Innovation Research Program, the Small Business Technology Transfer Program (STTR) differs in that the technical scope is limited, and offers must be from teams of small businesses and non-profit research institutions that will conduct joint research. The goal of the STTR Program is to transfer technology developed by universities and federal labs into the private marketplace through the entrepreneurship of a small business.

**Technology Transfer, Federal Laboratory Consortium for**
950 North Kings Highway, Suite 208
Cherry Hill, NJ 08034
Phone: 609-667-7727
Fax:  609-667-8009
www.federallabs.org

The Federal Laboratory Consortium for Technology Transfer (FLC) facilitates the rapid movement of federal laboratory research and technologies into the mainstream of the U.S. economy. The FLC network includes more than 800 federal laboratories and centers representing 17 different government agencies. It helps the private sector, state and local governments, and universities identify laboratory expertise and facilities that address needs for federally based products, processes and services.

## Technology Exchange Center (International)

P.O. Box 446
South Perth W.A., Australia 6151
Phone: +618-9367-7006
Fax: +618-9367-7343
www.technologyxchange.com

The International Technology Exchange Center (ITX) brings together buyers and sellers of technology, including those who want to expand their product markets. ITX has one of the largest databases of technology online, available for all members. ITX allows users to register company details and as many searches and technology entries as desired.

## Texas Centers for Border Economic Development including the Technology Transfer InfoSystem

The University of Texas
El Paso, TX 79968
Phone: 915-747-5000
www.utep.edu

The Texas Centers for Border Economic Development were established to provide economic and demographic information and analysis; conduct research that enhances economic development and provide technical assistance to business, industry and non-government organizations. The Centers enhance the economic position of the state by leading the development of the border region's economies. Their purpose is to stimulate economic development through the provisions of information, related research and technical assistance, policy development and analysis, outreach and education.

## Thomas Jefferson National Accelerator Facility

12000 Jefferson Avenue
Newport News, VA 23606
Phone: 757-269-7100

The Jefferson Lab's mission is to provide forefront scientific facilities, opportunities and leadership essential for discovering the fundamental nature of nuclear matter, to partner with industry to apply its advanced technology and to serve the nation and its communities through education and public outreach, all with uncompromising excellence in environment, health and safety.

## U.S. Department of Agriculture ARS Office of Technology Transfer

Willard Phelps
Office of Technology Transfer
Beltsville, MD 20705
Phone: 301-504-6905
Fax: 301-504-6060
http://ott.arsusda.gov

The Agricultural Research Service facilitates research in cooperation with private organizations, federal and state agencies looking to eliminate problems in the areas of plant and animal protection and production, and to promote the conservation and improvement of air, water, soil, farm products and human nutrition. The Technology Transfer Office helps business and industry locate the technology they need and facilitates and processes the moving of that technology from its source to potential users.

## U.S. Department of Transportation
Technology Sharing Program
Office of Research Policy & Technology Transfer
400 Seventh Street, S.W.
Washington, DC 20590
Phone: 202-366-4208
Fax:  202-366-3272
www.tsp.dot.gov

To increase the reliability, efficiency and cost effectiveness of transportation, as well as reduce its environmental impact, new advanced technologies are needed. The responsibility of the Department of Transportation's Technology Sharing Program is to ensure that the research is relevant to state and local government problems and the results of that research is available to those users.

## U.S. Geological Survey Technology Transfer Partnerships
John W. Powell Federal Building
12201 Sunrise Valley Drive
Reston, VA 20192
Phone: 703-648-4450
Fax:  703-648-4408
www.usgs.gov/tech-transfer/

The technology transfer program at the U.S. Geological Survey (USGC) is designed to leverage the research capabilities of USGS scientists with the commercial development potential of the private sector. The Technology Enterprise Office provides a link between the various research organizations within USGS and individuals or organizations in the private sector interested in access to USGS technology through patent licensing, Cooperative Research and Development Agreements (CRADAs) or other types of cooperative arrangements.

## Washington State Department of Transportation's Information Technology Transfer

Office of Technology Transfer
P.O. Box 47350
Olympia, WA 98504
Phone: 360-705-7802
Fax:  360-705-6823
www.wsdot.wa.gov/TA/T2Center/TechHp.html

Within the three sections of the Washington State Technology Transfer Center is the latest on local agency management and geographic information systems, including safety issues and opportunities to enhance technical management skills and knowledge of local agencies through the coordinated efforts in training, technical materials and technical advice.

## Wright-Patterson Air Force Base

Office of Air Force Technology Manager
AFMC/TTO-TTR
4375 Childlaw Road, Suite 6
Wright-Patterson AFB, OH 45433
Phone: 513-257-3998
Fax:    513-476-2138

**Wright-Patterson Air Force Base Technology Transition Office**
AF DUS&T Program
AFRL/XPTT
2310 8<sup>th</sup> Street
Bldg. 167, Room 126
Wright-Patterson AFB, OH 45433-7801
Phone: 937-656-9015
www.afrl.af.mil (Technology Transfer then Dual Use)

The Technology Transition Office is an information center that assists industry, academia and government in identifying Air Force technologies for transfer. It links business and industry with emerging technologies from the Air Force's four laboratories and assists in the progress of these technologies.

# about the authors

**Clifford M. Gross** received his Ph.D. from New York University. He served as Acting Director of the Graduate Program in Ergonomics and Biomechanics at New York University and Chairman of the Department of Biomechanics at New York Institute of Technology and founded and served as CEO of the Biomechanics Corp. of America. Dr. Gross was Research Professor and Director of the Center for Product Ergonomics at the University of South Florida. His laboratory was partially replicated in the Cooper-Hewitt National Design Museum of the Smithsonian Institution in March 1997 as part of a new exhibit on Henry Dryfuss and ergonomics. Dr. Gross holds eighteen patents and has produced numerous publications. His first book entitled, *The Right Fit*, published in 1996, describes how companies can increase market share and profitability using a biomechanics technology strategy. Dr. Gross is CEO and founder of UTEK Corporation, a U2B™ business development company dedicated to building a bridge between university research results and companies that can bring new technology to the marketplace.

**Uwe Reischl** received the bachelor's and master's degrees in Architecture from the University of California at Berkeley; he received a

Ph.D. degree in Environmental Health Sciences from the University of California at Berkeley and both a second Ph.D. in Occupational Medicine and an M.D. in General Medicine from the University of Ulm (Ulm, Germany). Dr. Reischl was an Assistant Professor at the University of California; Director of the Program in Industrial Health and Safety at Oakland University, Associate Professor, College of Public Health, University of South Florida and, most recently, a Scientific Advisor at the World Health Organization Center at the University of Ulm. He has fifteen years of experience in university teaching and research. Dr. Reischl is President of UTEK Corporation.

**Paul Abercrombie** is a veteran business journalist who has worked as a staff writer for several newspapers. While interviewing CEOs for the *Tampa Bay Business Journal*, Mr. Abercrombie became interested in how they use technology transfer to invigorate small technology companies. In addition to writing about technology themes, he has also contributed articles on a variety of topics to publications such as *Florida Trend* and *The Washington Post*. Mr. Abercrombie is President of Abercrombie Communications Inc.

# index

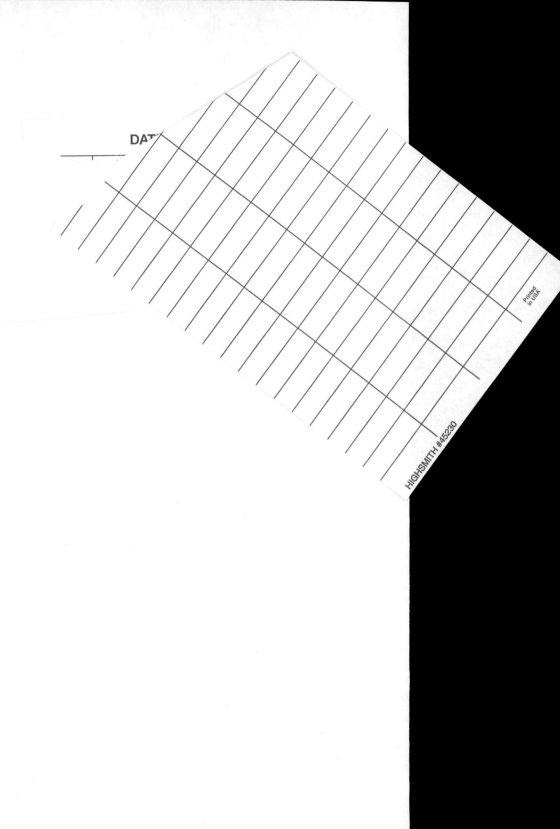

DAT

HIGHSMITH #45230

Printed in USA